RED CEDAR IN AUSTRALIA

RED CEDAR

IN AUSTRALIA

EDITED BY JOHN MCPHEE

HISTORIC HOUSES TRUST

Published by the Historic Houses Trust
of New South Wales
The Mint, 10 Macquarie Street
Sydney NSW 2000 Australia

Reprinted 2006

ISBN 1 876991 19 4

Edited by Vani Sripathy
Design by Bruce Smythe
Printed by Imago Productions, Singapore

Cover: *Brush scene, Brisbane Water* (detail), 1848
Conrad Martens
oil on canvas
Dixson Galleries,
State Library of New South Wales

Back cover: Sofa, c1840
maker unknown, Tasmania
red cedar, brass castors,
replacement horsehair upholstery
National Gallery of Australia

Gatefolds: Cedar (detail)
from R T Baker,
Cabinet Timbers of Australia,
published in Sydney in 1913
Private collection

Frontispiece: Door (detail) from an alderman's office, Sydney Town Hall, designed by David McBeath, 1878
Sydney Town Hall Collection,
City of Sydney
Photograph Brenton McGeachie

Margaret Betteridge is a Sydney heritage consultant and the Curator of the Sydney Town Hall Collection for the City of Sydney.

Clive Lucas is a practising architect, architectural historian and the author of books and articles on Australian architecture. His Sydney firm, Clive Lucas, Stapleton & Partners, is well known for its specialisation in building conservation.

David Mabberley is the Leidsuniversiteitsfonds Professor at the Nationaal Herbarium Nederland, University of Leiden, The Netherlands and an Honorary Research Associate for the Botanic Gardens Trust, Sydney.

John McPhee is a Sydney art historian and the Visiting Curator, Historic Houses Trust of New South Wales, responsible for curating *Red cedar in Australia.*

Rod Ritchie is the author of *Seeing the rainforests in nineteenth-century Australia* and various travel related books. He works as a web and print writer and publisher.

Ann Toy is Supervising Curator, Government House, Historic Houses Trust of New South Wales.

This book was published in association with the exhibition
Red cedar in Australia
held at the Museum of Sydney
on the site of first Government House
8 May to 15 August 2004

EXHIBITION TEAM

Curator
John McPhee

Text editor
Vani Sripathy

Exhibition Officer
Beth Hise

Exhibition Coordinator
Richard Taylor

Collections Management
Caroline Lorentz

Assistant Exhibition Officer
Joanna Gilmour

Exhibition Graphic Design
Bruce Smythe

Production Assistance
Christina Fedrigo
Beau Vandenberg

CONTENTS

FOREWORD

The many properties of the Historic Houses Trust of New South Wales are showcases for Australian red cedar. I never cease to marvel at the huge living specimen growing so luxuriantly in the service yard of Vaucluse House, almost challenging us to remove it before it completely engulfs the historic stone wood shed. At Elizabeth Bay House and Elizabeth Farm the polished cedar joinery gives warmth to their otherwise sophisticated and chaste interiors. And the history of Australian furniture design can be read through the collection of cedar furniture at Government House.

It is no wonder that the Trust was immediately interested when John McPhee, a leading expert on Australian decorative arts, proposed an exhibition and a book devoted to this singular tree.

Known as 'red gold', *Toona ciliata* became one of Australia's earliest exports and primary industries. Like the other and better known gold rushes, its promise of quick wealth lured opportunistic cedar-cutters. Their relentless pursuit of the tree through the coastal forests of New South Wales and Queensland led to its near extinction by the 1870s.

Red cedar in Australia is published in association with an exhibition of the same name held at the Museum of Sydney *on the site of first Government House* in mid 2004. Both exhibition and publication trace the discovery and classification of red cedar; its subsequent logging; its use in Australia's earliest houses, public buildings and furniture, as well as some unusual uses of the timber in boat building, grave boards, coffins, cigar boxes and assorted ephemera. Curated by John McPhee, the project draws on his exceptional knowledge of colonial decorative arts to which he has interwoven the social history of the timber-getters.

The discovery and use of this once abundant rainforest tree is a story of desperation and excess. Aside from its value as a useful building and joinery material, the beauty of cedar can be seen in many finely crafted objects and buildings from the 19th century – some of the best examples are included in the exhibition and have been gathered from around Australia. Some are being shown for the first time. John McPhee, together with contributing authors David Mabberley, Rod Ritchie, Clive Lucas, Ann Toy and Margaret Betteridge each bring their own scholarship to extend our understanding of this revered tree. To them, and to all who have loaned objects for the exhibition, the Trust is indebted.

Peter Watts
Director
Historic Houses Trust of New South Wales

Puzzle frame
1898
J A Dohnt,
Shepparton, Victoria
cedar, painted wood, glass

Private collection

Photograph
Brenton McGeachie

Made from 908 pieces of cigar boxes, the maker is shown on the left in the 1902 family portrait.

INTRODUCTION

JOHN MCPHEE

European colonists in New South Wales first saw red cedar in rainforest about 10 km from Rose Hill, now Parramatta, in December 1788. In a land where most timbers, particularly the tough eucalypts, were difficult to work with and blunted the colonists' poor quality tools, the more easily cut and cured cedar was recognised as a superior material. It soon became one of the colony's most sought after resources despite government attempts from as early as 1795 to control cedar felling.[1]

From the time red cedar was identified as a useful timber for the infant colony's immediate purposes of building, joinery and furniture making, its fate was sealed. The rapaciousness of the 19th-century forestry industry combined with land clearing for grazing destroyed vast tracts of the tree's rainforest habitat. By the 1870s cedar had become scarce in the rainforests of New South Wales and the cedar-getters moved into Queensland. Today the remaining trees are in remote rainforests or on land that has become part of a national park.

Throughout the 19th century cedar provided not just useful, but fashionable timber for extensive joinery in houses, commercial and public buildings as well as for the furniture that went into them. Its natural abundance, suitability for cabinetmaking and the colony's preference for dark polished timber that imitated the mahogany used in Europe led to such profligate use that by late in the century, an alternative wood had to be found. The shift in taste towards lighter interiors, timbers and painted joinery and the diminishing supply of cedar meant that it was seldom used for building in the 20th century.

Red cedar in Australia presents new research and assesses the history of *Toona ciliata*, the environment in which it grew, its subsequent destruction by the timber industry and the impact this had on both the Aboriginal and European population. It looks at the varied and sometimes unusual uses of red cedar in building and architecture, joinery and cabinetmaking, and the popular arts through the 19th and into

Australian red cedar
(*Toona ciliata*)

Photograph courtesy Kevin Diletti

Red cedar trees generally grow to a height of about 40 metres with a trunk diameter of up to 3 metres and have an open crown with large spreading limbs.

the 20th century, and considers the role that cedar might have played in late 19th-century nationalism.

In his elucidating chapter about the discovery and description of red cedar, botanical scholar David Mabberley discusses the many different trees that are known as cedar in different parts of the world. He describes the nature of cedar and the confusion that has surrounded its classification in the 19th and 20th centuries, with an informative account of the arguments among early botanists, some of whom wished to have Australian red cedar recognised as an endemic species, and the more recent recognition of its relationship with the toons of Asia, particularly India.

From 1791 and through the 19th century, samples of colonial timbers, including cedar, were sent to England for examination and evaluation. Some of the earliest assessments of cedar samples by English cabinetmakers considered it most suitable as a replacement for pine deal in the making of packing crates rather than for any fine cabinetmaking. But early colonists found a variety of uses for cedar. Its lightness, buoyancy, and resistance to insect infestation made it especially attractive to the colony's first boat builders, and it was later used in specialist boat building. Cedar was to also become the preferred timber for the New South Wales vice-regal rail carriage and for the panelling of Sydney's trams before their exteriors were painted in the 1930s. During the 19th century it was used in the construction of a wide variety of boxes for scientific instruments and even for some equipment.

Cross section of a cedar log
the annular rings marked with historical events from 1628–1918, 1924
Collection Powerhouse Museum, Sydney

Fascinated by large trees early 20th-century museum displays often emphasised the fact that many were much older than European habitation of Australia.

View of Sydney Cove, NSW
1804
Francis Jukes, engraver, after Edward Dayes and Thomas Watling
hand-coloured aquatint

The Beat Knoblauch Collection

Photograph Ross Heathcote

The cedar framework of the brig Portland, begun in 1797, can be seen in the colony's naval dockyard. Lack of shipwrights deferred the boat's completion until 1816.

Cedar's ability to take a high polish was undoubtedly the reason it was used for some of the colony's earliest coffins. Continuing their traditional role as undertakers, Sydney's cabinetmakers used cedar to make coffins for their wealthier clients who, in death, wished to keep up appearances. Weather resistant cedar grave boards presumably marked graves in remote districts before the visit of an itinerant stonemason, or were used by those unable to afford a more permanent memorial.

The fascination with the variety of the mostly unknown and unused Australian timbers persisted through the 19th and into the early 20th centuries and examples of colonial timbers were regularly exhibited in the Australian displays at the great colonial, inter-colonial and international exhibitions held in the 19th century. Cedar, with its original abundance, economic value (at least in the colony's eyes!), and beauty was among the most favoured of specimens at these exhibitions. Huge logs were regularly shown in the New South Wales and Queensland displays as examples of the potential wealth of primary industry. Among the more elegant exhibits were the timber 'books' (samples of individual timbers carved and painted to resemble books) made in the 1860s at the Pentridge Penal Establishment, Melbourne, under the direction of Colonel William Champ[2] and supplied by Baron von Mueller,[3] the Director of the Botanic Gardens, Melbourne, and an ardent believer in the economic potential of Australian timbers.

As well as their inclusion in numerous exhibitions, there were many articles and publications devoted to Australian timbers: the most well-known and authoritative being R T Baker's *Cabinet timbers of Australia*, published in 1913. In this book Baker gives a simple

***Cedar log shown by Cairns Chamber of Commerce at the Queensland International Exhibition of 1897**, from the Agricultural Journal, August 1897*

John Oxley Library, State Library of Queensland

Throughout the 19th century exhibition displays from New South Wales and Queensland emphasised the riches of colonial primary produce.

account of the historical use of Australian cedar, as well as a description of an extraordinarily wide variety of other timbers. But it is cedar for which he reserves a special place, acknowledging its importance historically and economically:

> *Foremost among such timbers is perhaps the 'Red Cedar', Cedrela toona, which has extended its range south from the Malay Peninsula into Queensland and New South Wales. Taken all in all, it is considered the best all-round timber in Australia, and is still in great demand for counters and internal woodworker decoration, although in recent years, for furniture making it has gone out of fashion.*[4]

Soon after cedar was discovered, the colonial administration, aware of its economic potential, tried to ensure that it remained the property of the government rather than sold for the benefit of enterprising settlers and timber-getters. Nevertheless cedar-getter contracts were eagerly sought and carefully issued and in the rush to locate, fell and transport cedar, enormous damage was done to the fragile ecosystems of the trees' rainforest habitats.

Rod Ritchie's account of the nature and beauty of the rainforest landscape in which cedar was found in New South Wales and Queensland is a poignant reminder of the losses incurred in the process of exploiting this timber. Ritchie writes of the ways in which cedar was located, felled, transported and milled. He describes the lasting impact of timber-getting and clearing land on the landscape, the devastation of the rainforest environment, and its subsequent impact on the Aboriginal population.

In the 1860s and 1870s artists and writers were some of the first to document the devastation done to the rainforests by timber-getters. Eugene von Guerard's coloured lithograph, *Cabbage tree forest, American Creek, NSW*, 1866–68,[5] after his own painting, carefully notes the original beauty of the rainforest while simultaneously depicting its destruction. The text accompanying the plate states:

> *This sylvan scene is situated at a distance of little more than ten miles from Wollongong, near the junction of a little stream upon which some prosaic devotee of the bottle has bestowed the dishonouring appellation of the Brandy and Water Creek, with the American Creek, and at the foot of a noble range of mountains. With the lofty bangalow palm, cabbage palm, the gigantic wild fig-tree, and the fire tree (otherwise known as the blaze tree) with its vividly scarlet blossoms, are intermingled the nettle tree, the rose-wood, the sassafras, the white-wood, the wild rose, numerous varieties of fern tree, and parasites innumerable; the whole thing being woven together into one dense and almost impenetrable mass of foliage. Unfortunately the progress of settlement is necessitating the destruction of some of these magnificent forests, which in many instances clothe a rich chocolate soil of especial value for the farmer. At the time this view was sketched, numerous fires had been kindled by the wood-cutters, and the stately giants were rapidly falling before the pitiless axe of the hardy pioneers of civilization.*

Rapturous descriptions of the rainforest were equalled by chilling accounts of the uncouth nature of itinerant workers. In an account of squatting and life in the bush the retired military officer and settler John Henderson wrote about the Illawarra district in which the red cedars grew, describing it as a 'thick mat of brushwood, woven together with vines, or creepers, of various kinds'.[6] Having set the scene, he describes the harsh conditions under which the cedar-getters worked and the difficulty experienced in locating, felling and transporting timber after the first easily procured trees were cut out. He focuses on the

Drue Walla, Illawarra
1840–44
Abraham Lincoln
pencil on paper
Mitchell Library, State Library of New South Wales

Lincoln observed that, 'this little Farmhouse is built almost entirely of cedar'. It shows the land clearance typical of 19th-century foresty and grazing practice.

28 timber specimens in the shape of books
c1866
Unidentified prisoner, Pentridge Penal Establishment, Melbourne
red cedar, and various other Australian timbers
Tasmanian Museum and Art Gallery

A scheme of prisoner reform activity resulted in these 'books' made of various Australian timbers.

nature of the timber-getters, their families, living conditions and their often squalid and intemperate lives. But while accusing the timber-getters of a profligate lifestyle, he also observes the inherent problems in a system which cut men off from contact with others, forced them to live far from the comforts of reasonable living and gave them occasional access to large quantities of money:

> *These sawyers and their mates are a strange wild set, comprising in general a good proportion of desperate ruffians, and sometimes a few runaways, they themselves commonly being ticket-of-leave men or emancipists. Two or three pair, accompanied by one or two men for falling, squaring small timber, and digging pits, shoulder their axes and saws, and with a sledge and dray-load of provisions, proceed to some solitary brush, where they make a little 'gunya', or hut, with a few sheets of bark, and commence operations. They labour very hard, stripping to the waist in the hottest summer days; but they live in extreme abundance, and, indeed, wastefulness, though their fare is but simple, consisting only of salt beef, damper, tea and sugar. From their migrating habits they are unable to have any kind of vegetables, but they invariably indulge in flour of the finest quality.*

Henderson notes that every two or three months the timber-getters would travel down the river for their wages, or 'settlement' as it was known. Their employers would charge excessively for rations and clothes and the timber-getters would:

> *... start off to the nearest public-house (perhaps a distance of forty miles), there to remain till they have spent every farthing, often exceeding thirty or forty pounds, when they return once more to the brush, in order to resume as before the same labour. They are certainly the most improvident set of men in the world ...*
>
> *In riding along some path leading through a brush, and bewildered and lost amid the various and endless mazes of cedar tracks, one will often stumble upon a miserable cabin, shut out from the genial rays of the sun, instinct with life in the shape of gigantic mosquitoes and other vermin, and inhabited by a lonely sawyer and his dirty and forbidding wife, or mistress, probably a ticket-of-leave woman or emancipist. If there are any children, which is occasionally the case, they are in the last stage of squalor and filth, their pale and emaciated features, already showing that fever and ague – the demons of these bushes – have begun to work with them.*

opposite
Brush scene, Brisbane Water
1848
Conrad Martens
oil on canvas
Dixson Galleries, State Library of New South Wales
Aboriginal pathways were often used by European explorers to pass through dense rainforest.

Chopping down a large tree in the primeval forest
from Eric Mjöberg, *Amongst the stoneage people in Queensland's wilderness*, published in Stockholm in 1918
Private collection
Aboriginal people would sometimes assist timber-getters in locating trees, particularly in Queensland where some were involved in timber-cutting operations.

Although Henderson writes about rum as the cause of the wildest behaviour in the camps, he lays the final blame at:

> *... the depraved and degraded class to which most of the sawyers belong; their loneliness and seclusion, being cut off during their whole time from any chance of good advice, or example, and the comparatively high pay for their work,*

together with the large sums which they receive at one time. The rafters are of the same class, and partake of the same habits.[7]

In 1839 Henderson had settled on the Upper Macleay River, in northern New South Wales, and the observations he made are undoubtedly those of experience. His writing reveals an awareness of the qualities of the Australian bush, the horrors of its convict past and the squalor of the hard life many were forced to live in the bush while earning a living, as well as the loneliness of many isolated settlers. Published in 1851, just as the gold rushes so dramatically changed the nature and composition of Australian society, Henderson's realistic depiction of colonial industry are unlike the later more idealised and romantic images of the landscape and pioneer life featured in the nationalism of writers like Henry Lawson[8] and painters of the Heidelberg School such as Fred McCubbin[9] and Tom Roberts.[10]

Henderson describes his own house at Elsineur, 184 km from Port Macquarie, as,

> *... built of long gum-slabs lined, made and partitioned with cedar boards, of which also the floor was made, and had a good fire-place faced with bricks, which had been carried all the way from Kempsey ... A verandah of cedar slabs, barked, and over grown with the fruit bearing passion vine.*

His book is illustrated with a charming image of himself seated in the interior of the cedar-lined house. A large simple dining table, a small side table, a set of hanging shelves, the fireplace surround, and perhaps even the chair the author sits in, are probably made by a farm worker or carpenter on the property who had some skills as a furniture maker.

Hutten der cederholzfaller (Blackall Range)
1897
J J Hogg, photographer
albumen silver photograph
John Oxley Library, State Library of Queensland (168491)
This Queensland cedar-getters' camp is more ordered and prosperous than those usually associated with the timber-getters' lifestyle.

***Interior of the author's house at Elsineur*,** from John Henderson's *Excursions and adventures in New South Wales …* published in London, 1851

Private collection

Henderson's house was lined and partitioned with cedar boards and had a cedar floor. The simple furniture was probably made by a farm worker or carpenter on the property.

Clive Lucas's chapter gives an account of the varying fashion for cedar in Australian houses. Used in colonial and late 19th and even some early 20th-century interiors, it imitated the European, particularly British, fashion for mahogany. Lucas also considers how cedar joinery and furniture was used in ecclesiastical, government and commercial interiors to create a sense of tradition and prosperity, particularly important for government and commercial enterprise in the 'boom and bust' economy of late 19th-century Australia. Importantly he points out the remarkable qualities of the unpainted cedar-lined interiors which survive in houses on some Queensland properties, as perhaps the most significant and peculiarly Australian use of timber. These have their origins in the almost makeshift interiors created by settlers such as Henderson.

Ann Toy and Margaret Betteridge both discuss the use of cedar in two of Sydney's most significant buildings, Government House and the Sydney Town Hall. In both buildings, a grand house and a public building, cedar was part of the overall plan of the interiors through several distinct periods. For over more than a century the furniture for both buildings was especially selected, designed, made for, and remains in, specific interiors. In the Town Hall, in particular, architectural carving plays a significant part in the enrichment of the interiors which were designed to celebrate the wealth of the colony's premier city.

Finally, my chapter on furniture and cabinetmaking gives an account of the use of cedar in furniture and objects made by carpenters, cabinetmakers, wood-carvers and craftspeople and its popularity with early furniture makers whatever their level of skill. The wide range of cedar furniture produced in

Newtown Post Office interior
1893
National Archives of Australia
(C4078, N20232)
Late in the 19th century polished cedar joinery complemented by cedar furniture was extensively used in government buildings to create a sense of tradition and prosperity.

the 19th century attests to the huge industry as well as continued fashion for the wood, at first among the rich and middle classes who wanted fine furniture for colonial mansions and houses to establish their success and position. Later, when the colony supported a larger importing trade and furniture regularly arrived from fashionable British suppliers, and taste for dark wood diminished, cedar was used extensively for traditional, not ostentatiously fashionable, furniture in commercial interiors such as banks, post offices and courthouses.

The book also raises some questions about the acceptance of the similarity of Indian and Australian cedar. Our recent awareness of the extent of trade between the British colonies of India and Australia has led to a re-evaluation of the possible origins of some colonial furniture.[11] This, perhaps more than anything else, points out the difficulties the furniture historian faces in establishing the origins, and sometimes the nature, of the timber used in colonial cabinetmaking. At times the chance of accurately determining the nature of timber that has been aged, perhaps stained, faded and polished over many decades, is almost impossible.

More puzzling is the existence of cigar boxes made from Australian cedar. By the mid 1880s New South Wales had become the main supplier of Australian grown tobacco and huge quantities of Australian grown and made cigars were sold. However in an effort to dispel the taint of home-grown manufacture most were boxed in locally made boxes labelled with Spanish and Cuban sounding names. In the late 19th century and well into the 20th there were a number of mills that specialised in providing cedar cut for cigar box manufacture. It has not been possible to distinguish any surviving boxes as those made in Australia from Australian red cedar.

The painters of the Heidelberg School used cedar cigar box lids, because of their convenient size, when painting out of doors.[12] Fellow painter Louis Abrahams,[13] whose family imported and packaged cigars, and the brother of the painter Aby Altson,[14] who was a city tobacconist, supplied the panels to them. Many of the paintings in the famous 9 x 5 impression exhibition of 1889[15] were painted on cigar box panels. Whether they are painted on panels taken from locally made or imported boxes has never been established.

Later in the century several of these nationalist painters made use of cedar panels to paint on. These had been recognised as a stable and conveniently light, pest-resistant support for oil painting early in the colony's artistic history. The convict artist W B Gould occasionally used cedar panels, probably originally intended or used for door panels in joinery or furniture, as the support for his paintings. The later artists, however, made use of cedar as a suitable background colour, sometimes leaving the timber exposed. Like the nationalist cabinetmakers, who found cedar a suitable wood for their work featuring Australian flora and fauna, these painters used cedar in a manner that celebrated the peculiar beauty of the wood.

As there is no longer a plentiful supply of Australian cedar for use in interiors and cabinetmaking, craft workers are among the few who occasionally make use of the now

Evans Patent Combination Van
model railway carriage, c1900
maker unknown, New South Wales
red cedar, pine, nickle-plated metal and cotton
Private collection
Photograph Brenton McGeachie

Electra
30" tonner, pond yacht on stand, 1880–1900
maker unknown, New South Wales
red cedar, kauri, Huon pine, western red cedar, cotton, brass
Collection: Jack Woodward, *Model Shipyard*, Balmain
Photograph Brenton McGeachie
Because of its ability to float and ease of working, cedar was a favourite timber of model boat builders.

almost legendary timber. Clive Lucas remarks on the fashion for exposing cedar panelling never intended to be seen which marked the revival of interest in historical properties in the mid 20th century.

A similar middle-class Australia claimed cedar for its own when, in the 1970s, the bicentennial celebration of Captain Cook's discovery of Australia inspired a new nationalism. Collectors sought out cedar furniture at auctions and in antique shops as something quintessentially Australian and part of their heritage. Their interests inspired an interior decorating style perceived as colonial and even reproduction furniture, usually made from imported timbers resembling cedar, is now common.

While the gum tree might be an almost ubiquitous symbol of the Australian landscape, it is cedar which has featured significantly in Australian building, architecture, interiors and cabinetmaking, as well as in the country's economic development. Cedar's beauty as a polished timber has earned it pride of place in the joinery and furniture of some of the finest interiors of the 19th century. Its use in some outback Queensland interiors has created some definitive Australian rooms. Used for purposes as diverse as coffin making and boat building, cedar's versatility made it popular with craft workers, but unsustained demand for the timber and its profligate use in the 19th century led to its rapid demise. Today it is largely through these surviving cedar objects that the qualities of this remarkable timber, and its remarkable tale, are remembered.

Still life, 1840
W B Gould
oil on panel (red cedar), pine (unidentified) frame
Private collection
Photograph
Ray Joyce

Painted on a cedar panel, probably originally made for use as a door panel in a piece of furniture, Gould's painting bears a label noting that it was framed by the well-known framemaker Robin Hood in Hobart Town, Van Diemen's Land.

Silent Bob Bates
1895
Tom Roberts
oil on panel
(red cedar)

Art Gallery of New South Wales

Photograph Jenni Carter for AGNSW

In exposing so much of the panel's surface, Roberts has shown an awareness of the prevailing aesthetic fashion as well as celebrating the colour and grain of the cedar panel.

1 EUROPEAN DISCOVERY, DESCRIPTION AND NAMING

DAVID MABBERLEY

Toona ciliata
(detail)
c1805–10
Ferdinand Bauer
pencil and
watercolour on paper

Plant nomenclature

Species names for plants are made up of two parts: an initial generic name denoting genus and a second specific name denoting species. A species is a group of individual plants or animals capable of interbreeding and producing fertile offspring and is usually given a common name – eg human, red cedar, rice. Closely related species belong to the same genus and therefore share the same generic name – eg *Toona ciliata*, *Toona sinensis*, *Toona sureni*. Similarly, closely related genera are grouped within a family – eg *Toona*, *Cedrela* and *Swietenia* are all genera in the family Meliaceae.

> *Dear Sir,*
>
> *Our settlement at Newcastle is going on extremely well under the direction of Lieut. Menzies ... we get plenty of very fine Coal & Cedar, which gives constant employment to our most disaffected Characters.*
>
> Philip Gidley King, Governor of New South Wales, Sydney, December 1804, to Sir Joseph Banks in England[1]

Red cedar[2] is the common Australian name for the species *Toona ciliata*, a rainforest tree of the mahogany family Meliaceae. Named after the qualities of its timber which resembles West Indian cedar (*Cedrela odorata*), a species of a closely related genus, red cedar also bears a close resemblance to the true mahoganies[3] of the genus *Swietenia*.

The name cedar is widely used in Australia for trees in the mahogany family, but in the northern hemisphere it more commonly refers to conifers related to the genus *Cedrus*, such as the cedars of Lebanon (*Cedrus libani*). This was the first use of the name, which derives from the

Forest, Illawarra, NSW, from Charles Wilkes (ed), *United States Exploring Expedition, 1838–1842*, published in London in 1845
J A Rolph, engraver, after Alfred Agate
coloured wood engraving
Courtesy Wollongong City Gallery
Agate's depiction of the rainforest concentrates on the great variety of the vegetation, including different kinds of palms and possibly cedars.

old French *cedre*, itself from the Greek *kedros* via Latin *cedrus*.[4] By the 17th century tropical trees with aromatic timber were also being called '*Cedrus*', notably the West Indian cedar, which in the 1690s was called '*Cedrus barbadensium* ...' by the British botanist and physician Leonard Plukenet. This connection was retained by the Irish naturalist and physician Patrick Browne when he named the West Indian cedar, *Cedrela*, in his 1756 publication *The Civil and Natural History of Jamaica in Three Parts*.

Meliaceae (the true mahogany family) in Australia

In Australia red cedar and white cedar (*Melia azedarach*) are just two of the 38 indigenous species representing 11 genera in the mahogany family.[5] Two of the genera (*Owenia*, the emu-apples, and *Synoum*, the bastard or scented rosewood) are restricted to Australia. Elsewhere the mahogany family comprises some 600 species in 51 genera throughout the tropics and subtropics, with a few extending to temperate regions. Many of these timbers[6] are important commercial commodities but natural stands of some have been severely depleted through over-exploitation. The original 'mahogany' used in Hepplewhite and Chippendale furniture was *Swietenia mahagoni*, a timber native to tropical America, which produced more graceful and woodworm-proof furniture than oak and walnut previously used in Europe. This species has suffered severe 'genetic erosion' due to the removal of the largest trees, leaving depauperate specimens as the only seed source. As a result, most 'mahogany' (if Meliaceous at all) seen today in Australia is derived from *Swietenia macrophylla,* which was introduced to Asia from tropical America in the 1870s and described as a new species from material cultivated in India. Today it is grown in plantations in Queensland and, as its name suggests, has larger leaves than *S. mahagoni*.

Besides red cedar and West Indian cedar, *Cedrela odorata*, other important timbers in world trade are *sapele* and *utile* (species of *Entandrophragma*), African mahogany (species of *Khaya*) and Nigeria golden walnut (*Lovoa trichilioides*) – all tall canopy trees from tropical Africa. Like red cedar and the true mahoganies, all of these have winged seeds that are released above the forest canopy and dispersed by the wind.

The other important Australian timbers in the mahogany family belong to *Dysoxylum* , a genus of some 80 species found throughout the tropical Asia–Pacific region. These have fleshy seeds that are dispersed by animals. The most familiar species is rosewood,[7] *Dysoxylum fraserianum*, which is restricted to southern Queensland and New South Wales, south to Wyong Shire. Familiar as a street tree in Sydney's eastern suburbs, it can grow into a large forest tree. The rose scented termite-resistant timber is used for furniture and carving.

Of the remaining 14 Australian *Dysoxylum* species, useful timbers include *Dysoxylum rufum* ('rusty mahogany' or bastard cedar pencil-wood of Queensland and New South Wales) used in cabinetwork and formerly joinery, *Dysoxylum pettigrewianum* ('scrub ironbark' or Cairns satinwood, of northern Queensland, New Guinea and the Solomon Islands) and *Dysoxylum mollissimum* (*D. muelleri*, the red bean or kedgy-kedgy of northern New South Wales, Queensland and the Malay Archipelago) both of which were also used in cabinetwork.

Distribution of Red Cedar

Toona ciliata

1 Pakistan
2 India
3 Nepal
4 Bhutan
5 Bangladesh
6 Burma
7 Thailand
8 Malaysia
9 Laos
10 Vietnam
11 Cambodia
12 China
13 Indonesia
14 Papua New Guinea
15 Australia

Banks (detail)
c1826
Charles Etienne Pierre Motte, after H Grevedon, 1826
lithograph

Rex Nan Kivell Collection, National Library of Australia

Banks's interest in the botany of the colony was closely associated with its commercial potential. Governor Phillip sent samples of cedar to Banks in 1791.

Red cedar in India: toon

> *A … travelling physician, who would wish to procure an* Arabian *or* Indian *plant, and without asking for it by its learned or vulgar name, should hunt for it in the woods by its* botanical character, *would resemble a geographer, who, desiring to find his way in a foreign city or province, should never inquire, by name, for a street or a town, but wait with his tables and instruments, for a proper occasion to determine its longitude and latitude.*
>
> Sir William Jones[8]

Toona ciliata is found growing wild from eastern Australia to New Britain, New Guinea, the Moluccas, northwards to southern China, India, Pakistan and Burma. In India it is known as *toon*, the origin of its first latin name, *Cedrela toona*, which, although coined by William Roxburgh, an 18th-century Scottish botanist working in India,

***Cedrela toona*,**
undated
artist unknown, India
pencil, pen and ink
and watercolour
Royal Botanic Gardens, Kew
Drawn for the British botanist William Roxburgh in India, this illustration, published as a coloured engraving in 1815, was the earliest printed representation of red cedar.

first appeared in print in Germany in 1803 in a letter from the Moravian missionary Johan Peter Rottler.[9] The Berlin botanist Carl Ludwig von Willdenow published the letter as *Botanische Bemerkungen auf der Hin-und Rückreise von Trankenbar nach Madras vom Herrn Missionair Rottler zu Trankenbar*[10] to which he added footnotes based on herbarium specimens that Rottler had sent him.

A few years later Captain Thomas Hardwicke could write:

> *Cedrela – the tree commonly called* Toon, *described by Sir WILLIAM JONES ... is found in the forests bordering the mountains below the ghauts* [sic]. *Grows to a tall tree, but seldom of considerable thickness. Is more in esteem for household furniture by* Europeans, *than for any use the natives put it to; bears resemblance to mahogany, but of much coarser fibre.*[11]

The timber of the tree had been used in shipbuilding by the British Navy before then, and its botanical identity had been accurately discussed by the polymath orientalist, Sir William Jones (1746–1794) – an account published posthumously in 1795 – though Jones did not give it a scientifically acceptable name.[12] But by this time Europeans had also found the tree in Australia and, reflecting its affinities, called it cedar.

At his own expense, Roxburgh[13] employed Indian artists to prepare watercolours (now held in Calcutta) of significant plants and directed that copies be made for the East India Company and others in England. A selection of these drawings, including a watercolour of *toon* by an unknown Indian artist, was chosen by Sir Joseph Banks[14] to become plates for a book.

Hawkesbury River view no 7, 1802
William Westall
pencil
National Library of Australia
The cedar samples sent to Banks in 1791 were gathered from the rainforest of the Hawkesbury River district.

Now preserved at the Royal Botanic Gardens, Kew, England, the drawing was the basis for the coloured engraving published in the second fascicle (1815) of Roxburgh's posthumously completed last volume of his three-volume work, *Plants of the Coast of Coromandel* (1795–1820) – the earliest published illustration of the plant. It was during the lengthy gestation of this book that Willdenow published his transcription of Rottler's letter, thereby validating for the first time Roxburgh's manuscript name:

> *Cedrela toona* Roxb[urgh].
>
> *Arbor grandis. Foliis pinnatis, 8-10-jugis, ovato-lanceolatis, integerrimus, undulatis, glabris. Flores in racemo composito terminale* [Tall tree with pinnate leaves having eight to ten pairs of ovate-lanceolate, entire, undulate, hairless leaflets. Flowers in a terminal compound raceme].

Willdenow added a helpful Latin footnote to allow readers to distinguish the new tree from the well-known West Indian cedar:

> *Cedrela odorata, calycibus corollisque nudis* [ie West Indian cedar, with hairless sepals and petals] *Cedrela toona, calycibus corollisque ciliatis* [ie *toon*, with sepals and petals with hairs round their margins].

Rottler's original (type) specimen, collected in Madras on 6 November 1799 by his fellow missionary, Johann Gottfried Klein, survives in Willdenow's personal herbarium, now preserved in the Botanisches Museum in Berlin with one of Roxburgh's *toon* collections. Roxburgh's own description[15] in his manuscript *Flora Indica* was not published until many years after his death. Being a doctor, he recorded the medicinal uses as well as the qualities of the timber:

A view of the River Hawkesbury, N S Wales, c1810
J W Lewin
watercolour
Dixson Galleries, State Library of New South Wales
In the earliest years of the 19th century the Hawkesbury River was the centre of the cedar-getters' trade.

> *The wood of this tree is much like mahogany, but lighter, consequently not so close in the grain. It is much used for furniture and various other purposes. The bark is a powerful astringent, and although not bitter is a tolerably good substitute for Peruvian Bark* [from species of *Cinchona*, the source of quinine] *in the cure of remitting and intermittent fevers; particularly when joined with a small portion of the powdered seed of Cesalpinia bonduccella* [sic, ie *Caesalpinia bonduc*, nicker bean] *(Kukufegei of the Bengalees), which is a most powerful bitter.*

In Asia today[16] the timber is used for building boats and houses, manufacturing furniture and carving tea chests, oilcasks, musical instruments and pencils. The flowers yield red and yellow dyes for silk, cotton and wool, and the bark is used to treat dysentery and dress wounds.

Red cedar in Australia: 'the mahogany of New Holland'

In December 1788, less than a year after the settlement of Sydney Cove by the First Fleet, red cedar was seen by Europeans in the 'North Brush'[17] some 10 km from the Rose Hill settlement, now Parramatta. It is likely that red cedar was the tree discussed in Governor Phillip's letter of 13 February 1790 to Thomas Townshend, First Viscount Sydney and Secretary of State in the Home Department in London:

> *The wood of some of these trees is very light; they are about the size of large walnut-trees, which they resemble; they shed their leaves, and bear a small fruit which is said to be very wholesome.*[18]

The 'North Brush' had been accessible since April 1788, allowing the tree's deciduousness,

Kiama, Illawarra, NSW, 1869
Henry Gritten,
after Robert Hoddle
after a sketch made
in 1830
oil on canvas
National Library of Australia

While not living there, the Aboriginal people of New South Wales made use of the rainforest's seasonal food and hunted the animals it sheltered. They made no known use of cedar.

an unusual thing for a native tree, to be seen during the intervening winter. It is likely that it was the colonists there who first used the name 'cedar' for the timber.[19]

Phillip himself saw the rainforests on the Hawkesbury in July 1789 when he must have encountered red cedar, though perhaps without recognising it. On 18 November 1791[20] he wrote to Evan Nepean, then Under-Secretary of State at the Home Department, about 'some specimens of the timber of this country ... put on board his Majesty's ship *Gorgon*'. These included pit-sawn planks of red cedar from the Hawkesbury, perhaps felled at Mangrove Creek or near present-day Windsor. The timber's resemblance to mahogany, *Swietenia mahagoni*, was to lead the Admiralty to order the empty convict transports to bring back red cedar and other timbers, first to the Cape, where there was a shortage, and then to the dockyards of England.[21]

As the Sydney settlement grew, red cedar was used to make the doors and window frames of new buildings, and to build colonial boats. Logs were carted overland to Parramatta or taken by boat down the Hawkesbury River through Broken Bay and south along the coast to Port Jackson. Its felling at such a rapid rate alarmed the authorities. On 2 April 1802 a general order was issued:[22]

> *It having been represented to the Governor that some of the Settlers at the Hawkesbury are making a traffic of the Cedar growing on or about that River, he strictly forbids any Cedar being cut down but by his particular permission to the Officer commanding at that place. And if any Cedar Logs or Planks are brought from any part of that River, to any other Settlement without the Governor's permission, such Logs or Planks will be seized for the purposes of the Government, and the Boats or Carts confiscated for public use.*

Red cedar was also found growing in abundance along the banks of the Hunter River where its presence was noted in Lieutenant James Grant's journal entry for 8 July 1801:[23]

> *... from the marks left on the trees, it should seem, that it* [the Hunter River] *is subject to be greatly overflowed at times, the cedar (or rather the mahogany of New Holland) growing near the river, appeared to have been immersed in water to the height of 40 or 50 feet.*[24]

In the same year a cedar camp was set up near present-day Maitland and a penal settlement established at the river's mouth. On 3 July a scarcely surprising proclamation was made:

The Governor judging it necessary for the public interest to declare the coals and timber which are to be procured at Hunter's River to be the exclusive property of the Crown, and having thought fit to establish a port at Freshwater Bay, within that view, he strictly forbids any boat or vessel going there for coal, timber, or any other purpose, without obtaining a special license from the Governor's Secretary.[25]

Controlling the exploitation and export of the Hunter's resources was seen as a way of raising tax to help defray the rising costs of maintaining the colony of New South Wales. A tax of one pound sterling was levied on 'every thousand feet square' of timber to be used in the colony, and two pounds for the same amount destined for export.[26] The penal settlement was abandoned but was re-founded in 1804 as King's Town, present-day Newcastle, under Lieutenant Charles Menzies,[27] appointed after his quelling of the Vinegar Hill convict 'rebellion' of 5 March that year.

Meanwhile a steady supply of red cedar poured into Sydney, the *Sydney Gazette* of 17 July 1803 noting:

CEDAR PLANK
THOMAS RABY begs leave to acquaint the Public that he has lately received a large Supply of BEAUTIFUL CEDAR from Hunter's River, which now lies open to Inspection and SALE, in Boards of all dimensions, adapted to any purpose, at his House on the Rocks.

The Irish settler Thomas Reibey ('Raby', 1769–1811) was the first non-military free settler to trade in Australia.[28] His cargo this time consisted of 55 logs 'of very rich cedar'.[29]

By 1800, the Illawarra was reached overland from Sydney but the abundant red cedar there and in the Shoalhaven was not to be officially recorded (from present-day Pig Island) until 1805[30] when the *Sydney Gazette* of 28 April carried another advertisement:

RICH CEDAR
Newly imported per the Governor Hunter, a quantity of fine and beautiful Cedar, adapted equally to Household furniture and building; and sawed into plank from half an inch to two and three inches in depth some of the sample exceeding two feet in breadth, six to eight feet long.

Gentlemen, Tradesmen, or others, wishing a large or small supply are requested to give indication thereof to I. Nicols, at the Hospital Wharf where the plank may be inspected.

The volumes of New South Wales red cedar being used in the town can be sensed from the *Sydney Gazette* of 11 May 1811, where it was announced, 'The Contractors for Building the General Hospital[31] are ready to treat with any Persons who will engage to supply 100,000 Feet of Sawn Timber of Sizes and 50,000 Feet of Sawn Cedar'. That year the first cargo from the Illawarra was reported.[32] The red cedar populations in what is now Queensland were to be exploited only much later. In 1829, though, the Scottish journalist Robert Mudie could already write prophetically of the Illawarra in his anonymous *Picture of Australia* (pp132–3), that red cedar was the 'most esteemed timber ... and will probably continue so till the forest be exhausted'.

George Caley: the first scientific description of red cedar in Australia

You will find at Sydney a young man Geo. Caley whom I have sent out to collect plants for me [H]*e is zealous & active & if he has not connected himself improperly there is likely to make a useful collection.*

Sir Joseph Banks[33] to Robert Brown, about to leave England as Banks's appointee as naturalist on Matthew Flinders's *Investigator* voyage (1801–05)

Although red cedar was so well known to the early colonists, there was no competent botanist in the early settlement to settle the tree's botanical identity. No specimens of red cedar, other than of wood, seem to have been gathered in Australia until Sir Joseph Banks's collector, the Yorkshireman George Caley (1770–1829), began forming a herbarium and acquired sprigs of cedar.

Cedar
from R T Baker, *Cabinet timbers of Australia*, published in Sydney in 1913
Private collection
Baker's book celebrated the potential of many Australian timbers, including cedar, which he noted as being out of fashion early in the 20th century.

George Caley,[34] always rather a difficult employee, had arrived in Sydney on 30 March 1800 aboard the *Speedy* with the new governor Philip Gidley King. He settled in Parramatta where he built up a herbarium and other natural history collections, including a botanic garden. With the appointment of Menzies as commandant to the new settlement at the mouth of the Hunter River, came an opportunity for Caley to extend his botanical domain while Flinders's naturalist Robert Brown was away working in Tasmania. Menzies's orders from King included the following:[35]

With these Instructions you will receive my appointment for commanding and superintending the settlement to be formed at the Coal Harbour and Hunter's River ...

After having taken the necessary measures for securing the stores, provisions, yourself, and people, you will cause the prisoners to be employed in getting as many coals as possible at hand, and procuring cedar from the upper part of the harbour.

The hours of the convicts labour is to be from daylight till eight o'clock, from half-past eight till noon, and from half-past one till sunset.

The convicts being servants of the Crown till the time for which they are sentenced expires, their labour is to be for the public; and you are to take particular notice of the good or bad behaviour, that proper notice may be taken of them hereafter.

You are not to suffer any individuals to cut cedar, dig or load coals, but by my particular permit in writing.

You are to take care that no private vessel loads with coals or timber obtained by those at public labour.

If any vessel comes to the river without my license, you will confine the crew and detain the vessel (by scuttling her) until you receive my orders.

So Caley was dispatched as part of the 'Expedition to Newcastle', as the *Sydney Gazette* of 1 April 1804 called it. He set sail from Sydney on the *James*, one of the three boats owned by Thomas Reibey, the coal, wheat and timber merchant,[36] arriving at the Hunter on 28 March and was among the founders of what is today's Newcastle. Brown's natural history artist Ferdinand Bauer travelled with Menzies on the *Lady Nelson*. On the journey Menzies foiled a plot by the convicts to overrun his party but, despite these tensions,[37] Caley and Bauer were able to work, collecting and drawing. Shortly after arriving at the settlement Bauer made the first field drawing (of fruiting material) of red cedar in Australia which was eventually to be worked up into a finished watercolour for the British Admiralty.[38] It was too late in the season for him to draw flowers.

However Caley must already have seen flowering material because he was able to write a description of it, probably at Parramatta, before the next season. Caley often delayed writing up descriptions[39] in his notebook so, unlike parts of Robert Brown's diary,[40] it cannot be considered a field notebook. The red cedar descriptions among Caley's Hunter River plants were likely to have been from an expedition during the previous Spring, but collected where or by whom has yet to be ascertained. It may have been collected in the Parramatta area by one of the Aborigines Caley employed to assist him.[41]

But Caley could also have received it from a contact like Reibey[42] visiting the Hawkesbury, or even the Hunter itself. A sheet of the flowering material labelled as being from the Hunter and '1804 [ie likely the year sent]', was to be preserved in Banks's personal herbarium, later forming the Banksian Department in the British Museum, now part of the herbarium of the Department of Botany in The Natural History Museum, South Kensington, London. It is still there, and is most likely to be the oldest surviving red cedar specimen from Australia.

Caley not only had what was probably the first collection of red cedar flowers in the country, he also drew up the first known botanical description[43] of the plant here. With terminology – esoteric even for his time – reflecting his lack of a formal training in botany, he wrote it in one of his botanical notebooks now preserved in The Natural History Museum, London:

A bivouac, day break on the Illawarra Mountains, 1827
Augustus Earle
watercolour
Rex Nan Kivell Collection
National Library of Australia
Travelling through dense rainforest with its huge trees was a novelty for many European explorers and travellers who were amazed at the size of some of the trees.

Cedar

Cal[yx] *Cup*[ular] *1 leaf, very small, with 5 divisions, of a greenish white color, segments blunt, being close to the blossom* [= corolla].

Bloss[om, ie corolla]. *Petals 5, unequal, somewhat egg oblong, a little bent inwards at the end, hairy at the edge, erect, of a dullish white color.*

Stam[ens]. *Filaments 5, white, flat, broadest at the base, tapering upwards, hairy, shorter than the petals, fixed to the nectary. Approaching anthers*

Toona ciliata, collected on the Hunter River by Robert Brown, 22 October 1804
plant specimen mounted on paper, pen & ink, printed paper labels

Specimens of flowering cedar gathered by Brown were used by the botanical artist Ferdinand Bauer to complete his watercolour.

incumbent, double, broad, of a dark brown color, nectaries 5, orange colored bodies fixed to the base of the germen [= ovary], *with white hairs at the top – petals with a tinge of green at the ends.*

S[eed]. *Vess*[el, ie capsule] *not ripe, top shaped, with 5 cells. Leaves* [= seeds] *winged, of a brownish green on the upper side, shining, but of a lighter green underneath and shining.*

Early the following Spring before red cedar was again in flower, Bauer's leader and Flinders's naturalist, the Scottish botanist Robert Brown (1773–1858),[44] went to visit Caley at Parramatta to discuss the new specimens that had been found in his absence, in particular, red cedar. On 14 September Brown wrote in his diary:[45]

Cedar in flower ... It is not a Melia [ie in the same genus as *Melia azedarach*, white cedar] *but according to Mr Caley Pentandrous* [ie the flowers with five stamens; those of white cedar have eight or ten].

Clearly Caley did not give him a specimen, so that Brown could not check this for himself. Brown was anxious to see the 'Coal River' and arrived at King's Town on 12 October, noting next day:[46]

Mr Menzies has already built a very sufficient Stone w[h]*arf, along side of which small vessels may take in their cargoes, whether coals or cedar.*

Brown himself saw the cedar-getters on the Hunter on 19 October, 'about 4 or 5 miles below mount Anna', between present-day Maitland and Morpeth on the Hunter River,[47] and two days later remarked on the 'Cedar rather sparingly & particularly on the bank'.[48] On 22 October he collected flowering material at the junction of the Hunter and Williams Rivers[49] which Bauer sketched and added to his completed watercolour showing both flowers and fruits, now preserved in The Natural History Museum in London.[50]

Brown called it '*Cedrella* [sic] *orientalis*' in his manuscripts, but in a letter of 21 February 1805 sent to Banks from Sydney[51] he wrote:

The small box which I now send contains the few seeds collected [on the Rivers Grose and Hawkesbury]. *To fill up the box I have added a very few specimens of such of my latest acquisitions, as either from their novelty or singularity I thought would be most interesting ... The cedar of the colony, with flowers and capsules to shew its affinity to Cedrela, from which, however, it differs in some points.*

Toona ciliata, collected in the Hunter Valley by George Caley, c1804 plant specimen mounted on paper, pen & ink, printed paper labels

© The Natural History Museum, London

This is probably the oldest surviving red cedar specimen collected in Australia.

Indeed, Brown, probably the greatest botanist of his age and certainly the greatest ever to visit Australia, was right, though scientists did not take his observations into account for many

Robert Brown,
1864
Henry William Pickersgill
oil on canvas
National Library of Australia
The Scottish botanist Robert Brown was the naturalist attached to Matthew Flinders's scientific expedition to explore and map the coastline of Australia.

years, the tree being referred to as *Cedrela* until well into the 20th century.

It has now been shown that the Old World species of '*Cedrela*' are quite distinct from those of the Americas in having seeds generally with wings at both ends or, if only at one end, the seed attached to the lower part of the middle of the fruit. In American *Cedrela* the seeds are at only one end and attached near the top of the middle of the fruit. The Old World trees now bear the Latin and Indian vernacular name, *Toona*. That name was published firstly as a section of the genus *Cedrela*, but raised to the rank of a genus by the Bavarian judge and botanist Max Roemer in 1846.[52] It was Roemer who coined the currently accepted name, *Toona ciliata*, picking up the species name from the word in Willdenow's original 1803 description of the hairy margins of the sepals and petals.

The genus Toona

Palaeontologists have referred a number of fossils from the Tertiary Period to the modern genus *Toona*. These include *Toona sulcata* from the Eocene London Clay[53] though that fossil was originally described in 1840 as *Cupressinites sulcatus*, indicating that the author thought it was a cypress-like conifer, resembling modern species of *Cupressus*! Many others have been found in Europe. From the Tertiary deposits at Wingello, near Bundanoon in the Southern Highlands of New South Wales, Henry Deane described a leaflet he took to be allied to *Toona* ('*Cedrela*'), naming it *Cedrelophyllum antiqua* [sic] in 1902.[54]

There are four to five living species of *Toona* recognised today and they are all now centred on south-east Asia and the Malay Archipelago. Probably the first known to Europeans was the *surian* or *suren*, *Toona sureni*, recorded by Georg Eberhard Rumpf (Rumphius) on Ambon in the Moluccas, Indonesia, in the 17th century and illustrated for his monumental seven-volume *Herbarium amboinense* (1741–55). This book includes probably the first European representation of any species of *Toona*, a plate published long after Rumphius's death. It is remarkable[55] that his work survived to be published in The Netherlands at all because of a series of personal disasters. Rumphius was struck blind at the age of 42 and all his surviving writings were prepared with the help of others. Then his original drawings and most of his original manuscripts and natural history collections were all lost in a fire in 1687. The first copy of his manuscript for the book was later lost in a ship sunk by the French but a second eventually got through, bringing the first European description for publication in 1743. He wrote of *Toona sureni*:[56]

Toona sureni,
from G E Rumphius,
Herbarium Amboinense,
vol 3, published in Amsterdam in 1743
engraving

Botanic Gardens Trust Library, Sydney

This is probably the first European representation of any species of Toona.

Mullet Creek, Illawarra, 1853
Conrad Martens
watercolour, gouache, gum arabic, pencil
Wollongong City Gallery
The George and Nerissa Johnson Memorial Bequest, purchased 1992
The huge trees of the rainforest: cedars and figs, and other exotic vegetation, fascinated artists and were a popular subject among their patrons.

Suren is a tall and straight tree, with wings [= buttresses] *near the roots, where the trunk is a fathom round, with a smooth, dry and very tough bark … The fruits are oblong berries, like Olive pits or a peeled Gnemon* [= Gnetum gnemon] *fruit, but smaller … which open up into five parts, showing a pentagonal light kernel or pillar inside, covered by five wings, of a light and spongy substance, and inside each little cell between the wings are one or two thin little scales, with a bulge in the centre, wherein lies a small kernel, like Sesamum* [= sesame], *which is the seed …* [The local people] *are very fond of this wood, and make light proas from it, which are light, do not tear, and are durable in salt water. When they find thick trunks, they make planks and ribs for their important houses.*

Suren differs from red cedar most obviously in its clearly bumpy excrescences on the fruits, the fruits in red cedar being smooth while the pistils inside the flowers are densely hairy compared with red cedar where they are hairless.[57] The bark is of medicinal importance locally and leaf extracts have an antibiotic activity effective against *Staphylococcus*.

Of the known *Toona* species, red cedar is the only one to reach as far south as Australia in the wild state, but occasionally Chinese mahogany (Ch'uen tien shu), the montane *T. sinensis*, native from India eastwards to peninsular Malaysia,[58] is seen in Australian gardens. It is readily distinguishable from red cedar by the toothed margins of its leaves and its 'obnoxiously pungent' bark. It is a much hardier plant and can be grown outside, even in Europe, where it is a conspicuous pollution-tolerant avenue tree to be seen in parts of central Paris for example. It is used for coffee shade in many tropical countries and the leaves and young shoots are eaten as a vegetable in Malaysia and China. The bark is widely used as an astringent, the powdered root as a diuretic and the fruits for treating eye infections. In Australasia a form of Chinese mahogany with brilliant pink spring foliage turning cream before becoming green, has been selected and is available as a cultivar, known as 'Flamingo'.[59]

Toona ciliata

Red cedar trees in Australia[60] are deciduous, generally growing to a height of about 40 m with a trunk diameter of up to 3 m, though much larger specimens have been recorded. Older trees have large buttresses extending well up the trunk. The scaly bark is greyish or brown and falls off as oblong or irregular pieces, leaving shallow depressions. Inside, the bark is chocolate brown and the living tissue is pinkish, red, and fragrant, but bitter in taste. The crown is open with large spreading limbs. The leaves are up to 50 cm long and have 8 to 20 leaflets that are reddish when young.

The trees are monoecious, that is, with separate male and female flowers, in this case in the same inflorescence (thyrsus). The fragrant insect-pollinated flowers are borne in large branched inflorescences (thyrsi) to 50 cm long arising from the axils of the leaves. Usually successive waves of male and then female flowers open from October to November, with the males far outnumbering females. The white petals are some 4 to 5 mm long and each, as Caley noted, has only five stamens. The fruit capsule is up to 26 mm long and splits into five chambers, each containing four to seven seeds, winged at both ends. The seeds are released between January and March and 550,000 are needed to make up a kilogram.

For many years, botanists in Europe considered, correctly, that the Indian, Indonesian and Australian plants of *Toona ciliata* (*Cedrela toona*) all belonged to a single widespread species, but in the middle of the 19th century, Australian botanists argued that populations of the tree here belonged to a distinct, but closely allied, species which they called *Cedrela*,[61] later *Toona, australis*. There

Brush Scene, Illawarra
from *Sketches in the environs of Sydney*, 1850
Conrad Martens
lithograph
Mitchell Library, State Library of New South Wales
As travel in the rainforest was uncomfortable, city dwellers were pleased to be shown glimpses as portrayed by artists.

are indeed some minor geographic differences across the species as a whole. In Asia, in the north-western part of its range, *Toona ciliata* is marked by its hairless stamens. Further research may show that it could be useful to recognise those with hairy stamens, the other populations, wild from Thailand southwards to Australia, as a distinct geographical subspecies.[62]

Cedar tip moth
(Hypsipyla robusta)
Paul Ovenden, Nature Focus, Australian Museum
Laying its eggs in the tips of growing shoots and causing the trees to branch out, cedar tip moths have been the main reason commercially successful plantations of cedar have not been established in Australia.

By comparison with the range of *Toona* fossils, the distribution is now much reduced. The main centre in Australia is from Gympie, Queensland, south to Ulladulla, New South Wales. Disjunct Queensland populations are on the Eungella Range west of Mackay, around the Atherton area and near Coen, with the northernmost on the Claudie River. Its altitudinal range is from near sea-level to about 1100 m. It was formerly plentiful on the alluvial flats of the larger rivers of New South Wales where trees to 60 m tall, that is as tall as the Queen Victoria Building in George Street, Sydney, were said to grow, a confirmed record being a tree reaching 54.5 m.

Red cedar grows best on rich volcanic or alluvial soils. In the 'Big Scrub', Dorrigo, and Levers Plateau, the Border Ranges and McPherson Ranges it grows on soils derived from basalt. Its latitudinal range in Australia means that it can grow in tropical and subtropical rainforests to warm temperate ones. Characteristic tree species associated with it are booyong (species of *Argyrodendron*), sassafras (*Doryphora sassafras*) and occasionally hoop pine (*Araucaria cunninghamii*). Plantations of red cedar generally fail in Australia as well as in the Solomon Islands and Fiji, because of the depredations of the cedar tip moth, whose larvae burrow into the growing-tips as well as the flowers and fruits. In Hawaii, however, red cedar from Australia is one of the most important plantation timbers and has even become weedy there[63] as it is in southern Africa.[64]

Because of intensive exploitation, the remaining stands in Australia are largely in scattered remote localities in New South Wales, and on the Eungella Range and the Atherton Tableland in Queensland. The most readily accessible large red cedar in Australia today is probably a tree in the Gadgarra State Forest near Lake Eacham on the Atherton Tableland. It is 35 m tall and has a clear bole to 18 m, with a diameter of some 7 m.

The author wishes to acknowledge the assistance of the following people in the preparation of this essay: Dr E M Beekman, University of Massachusetts, Amherst; Doug Benson and Miguel Garcia, Botanic Gardens Trust, Sydney; Neil Chambers, Joseph Banks Archive Project, The Natural History Museum, London; Dr Jennifer Edmonds, University of Leeds, UK; John Flanagan and Marilyn Ward, The Library, Royal Botanic Gardens, Kew; Professor H Walter Lack, Botanisches Museum, Berlin; Judith Magee, Botany Library, The Natural History Museum, London; Joan Webb, North Ryde.

Toona ciliata
c1805–10
Ferdinand Bauer
pencil and watercolour on paper

Bauer's finely detailed watercolour was painted from plant material collected at the Hunter River, New South Wales, in April and October 1804.

2

THE RED CEDAR TIMBER INDUSTRY IN NEW SOUTH WALES AND QUEENSLAND

ROD RITCHIE

On arrival in Sydney Cove in 1788, the new settlers were confronted by low, stunted eucalypts growing on the sandstone plateau of the harbour foreshores. The trees proved difficult to saw for the timber badly needed for building the penal colony, particularly with the poor quality tools the colonists had been sent with. The extremely hard eucalypt timbers were not only laborious to cut, but when used unseasoned, would often twist, split and shrink.

So initially, the colonists used the plentiful black wattles to build the foundations for their 'wattle-and-daub' buildings, packing clay onto sapling frames to construct the walls, and using the saplings for rafters. The native casuarinas split easily for roof shingles and made reasonable, if short-lasting, framing timber. In time, exploratory parties found good quality eucalypts in the sheltered valleys of northern and western Sydney and once hard, sharp tools arrived, this new resource was felled and sawn into timber for building frames, or adze-squared for industrial purposes such as yards, piers and wharves. Convict labour was used for the exhausting work of felling and sawing the trees into usable timber.

After brick kilns were built and sandstone quarries established, the new settlers were keen to find trees that would provide the type of softwood joinery timbers they were familiar with in Europe. Soon after settlement, they discovered red cedar on the banks of the Hawkesbury River. Its durable, easily seasoned wood was ideal for joinery and furniture. The deep red heartwood from the largest trees was particularly prized for cabinetmaking, especially if it was dark and heavily figured. Unlike eucalypts, red cedar logs floated and since roads were scarce, could be transported to market via the waterways.

Governor Phillip sent samples of Hawkesbury red cedar back to England in 1791 along with potted native plants destined for the naturalist Joseph Banks.[1] As the economic and utilitarian value of the timber became recognised and the demand for housing and

A splitters' camp on the upper Clarence River
1869
J W Lindt
albumen silver photograph
Richmond River Historical Society Inc
Studio photographs of people engaged in typical pioneering occupations were popular among photographers and painters in the late 19th century and part of a nationalist movement.

public buildings for the settlement increased, Hawkesbury district cedar was sent to Sydney in large quantities, either by cart from Parramatta, or by vessel from Broken Bay. By 1795 the local trade in cedar was so great that the colonial government, realising that it was losing control of the fledgling industry, issued a regulation limiting cutting. Despite this, a shipment of 60 large cedar logs was sent that year to Madras in India for use in the British naval depot there. By 1802, further government regulation was considered necessary and Governor King issued a general order banning cedar cutting without authority. Illegally cut cedar was to be confiscated, along with any vessel carrying it. However, there is no evidence that government regulation was effective in stemming the trade in red cedar. Cedar cutting was seen as a way to make quick money at a time when any possible opportunity to get ahead was taken.

Rafting cedar from *The Town and Country Journal*, December 1881

John Oxley Library, State Library of Queensland (21247)

Because it floated and was most easily transported out of the rainforest by river, log rafts were often used to take cedar to sawmills.

Reasonably abundant, red cedar formed the basis of the New South Wales 'timber rush' of the early 19th century and, along with whaling, became an important economic base for the young colony. Red cedar was found from Ulladulla on the far south coast of New South Wales to the Daintree River north of Cairns. It grew in the rainforests that covered parts of the alluvial coastal plains of the major river systems and also in the moderately fertile, high rainfall sections of the Great Dividing Range that were free from the influences of fire. As European settlement expanded, and land and sea expeditions reported the vegetation types in new areas, it wasn't long before people realised that huge quantities of red cedar were there for the taking. Spurred on by the promise of quick returns, timber-cutters began searching unclaimed areas well in advance of those looking for new grazing and farming lands. They were not in competition with squatters or small selectors since they required no land tenure. Demand for the timber grew, both from increased local consumption and from a flourishing export trade, and in little over 20 years from settlement, sawn timber became the colony's third largest export. By 1836 New South Wales was annually exporting cedar valued at £14,000.

After the Hawkesbury, the Illawarra was the next region to be logged for cedar, followed by the rainforests of the Shoalhaven. The Hunter River, found by Lieutenant Shortland in 1797, had cedar-cutters working around present-day Maitland in 1801. When the convict settlement of Newcastle was established a few years later, prisoners were employed in timber cutting and coal mining. The river was the perfect conduit for transporting the logs and cutters worked the

Cedar (Cedrela toona), portion of a giant tree – age 350 years
1924
photographer unknown, Sydney
gelatin silver photograph

Collection Powerhouse Museum, Sydney

The huge size of some cedars continued to fascinate the 20th-century public. A slab and a cross section from this specimen were presented to the Museum of Applied Arts and Sciences, now the Powerhouse Museum, Sydney.

riverbank forests for the next 20 years. By 1829 cutters had arrived at the Manning River and three years later, the Macleay River. By 1838 the Clarence River had been reached, then the big prize, four years later, the 780 km^2 stretch of land known as the 'Big Scrub', encompassing the rainforests of the Richmond River district. Timber cutting in the Tweed River, accessed by sea from southern ports, began in the 1840s. By the 1870s cedar was so scarce on the lowlands and the upper reaches of the rivers in New South Wales that cutters began scouring for the tree in the more inaccessible mountain ranges.

The southern Queensland forests around the convict settlement of Moreton Bay were first cut in 1828 when 'government' timber gangs took local cedar for building purposes. Aborigines fiercely protected their lands and during the 1830s regularly attacked timber-cutting parties. By 1847 nearly all the accessible cedar had been cut in south-eastern Queensland. In 1859, when Queensland was declared a separate colony, cedar from Wide Bay was being exported to New South Wales. In the 1860s the new government regulated the timber industry, issuing timber-cutting licences. Logs had to be removed from their felling location within three months and trees within 3.2 km of railway lines were protected. By this time activity centred on north Queensland

Railway carriages used for timber-getting, Kipper Creek, Queensland
1920s
photographer unknown

John Oxley Library, State Library of Queensland (11778)

The value of cedar made the building of elaborate methods to bring them out of the forests worthwhile.

where the tropical rainforests held vast quantities of valuable timber species. On the Johnstone, Mossman and Daintree rivers, cutters were hard at work in the 1870s. Logs were floated to river mouths, then loaded onto ships for transport south. Speculators from Melbourne were engaged in at least one operation, and the *Queenslander* newspaper noted:

> *No Queenslander receives any part of the profits of the trade ... When the timber camps on the Daintree are abandoned, there will be no settlers left behind. We have in fact sold the magnificent cedar forests on the northern rivers to Victorian speculators for the paltry sum received as licence fees ...*[2]

By the early 1880s, when the north Queensland coastal rivers were denuded of their cedar, a timber rush began on the Atherton Tableland rainforests, Australia's last great rainforest massif to be exploited. Access was made easier for the cutters by the activity of tin miners who were forming tracks through the forest. As the government licensing system did not allocate timber rights, ownership was claimed by falling trees. Before long valuable rainforest logs were piling up with little chance of ever reaching a market because there were no transport routes to the coast.[3] When settlers were allowed to select land on the tableland at the turn of the century, they embarked on a huge clearing task of an area that included many valuable timber species such as rosewood, white beech and silky oak. In another lamentable exercise, because timber dealers were only really interested in paying fair prices for the remaining red cedar, countless valuable trees were simply felled and burnt in the huge conflagrations that were common as land was cleared for agriculture.

In the late 19th century, governments realised that forests needed protection from agricultural clearing and poor timber harvesting methods and in 1871, set up a forestry branch within the New South Wales Lands Department to manage production in a more orderly fashion. However it was not until 1909, after a Royal Commission into the state of the timber industry, that a separate Forestry Department was established. As the Commissioner noted: 'Probably no section of business under Government control has experienced greater vicissitudes in its management or less consideration than that connected with our forests'.[4] In Queensland, a forestry branch was created in 1900, within the Department of Public Lands. During the 20th century, with

Bowman township, near Gloucester, NSW, c1900
photographer unknown,
New South Wales
print-out paper photograph
Josef Lebovic Gallery

Remote settlements of timber-getters were often the origin of new towns which grew up in the land cleared of vegetation and provided services for the industry.

most of the original red cedar having been largely cut out, rural landholders often traded in relict trees. Many trees remained in inaccessible locations to await the advent of crawler tractors and chainsaws in the mid 20th century.

Felling the trees

Finding the deciduous cedar trees was relatively easy – in spring the blush of their copper-red new foliage contrasted with their deep green, rainforested surroundings. The cedars were particularly evident from a high vantage point and when an area was discovered, no time was wasted in exploiting the resource. Alexander Harris, who worked as a timber-cutter in the Illawarra in the 1830s, recorded the fate of the region's rich forests that he helped deplete:

> *... nearly a hundred pair of sawyers had gradually come down from Sydney and gathered into this mountain, and were (as they continued to be for years afterwards) slaughtering away in all directions, it will not be wondered at that the pride of Five Islands Cedar Brush is long since gone ...*[5]

Ironically, by the time Eugene von Guerard's lithograph *Cabbage tree forest, American Creek* appeared in 1867, the American Creek rainforests in the Illawarra had long been logged and cleared for agriculture.

Life in the forest for the timber-cutters was hard. Frequent wet weather made working conditions dangerous and provisions were difficult to come by. Cutters lived by their wits and carried a gun for hunting and for protection from Aborigines. A journalist for the *Sydney Morning Herald* described a team of cutters from

Ballina after they had picked up their cheques on completing a 12-month stint in the bush:

> *... the years of toil hardly suffered to provide such a debauch as the bushman looked for as a reward for his labours. The rum cask was his Bible, and the grossest scenes of vice his ambition.*[6]

Not often mentioned was the fact that timber licences didn't allow for permanent settlements. Cutters moved around, sometimes with their family, making the best shelter that was possible under the circumstances. The dangerous nature of their work, the distance from medical care and the isolation were real drawbacks to leading a normal life. Alexander Harris noticed that the Illawarra cedar-getters, whose work deep in the rainforests meant they saw little light, were distinguishable by their pale colour: 'A few months residence and hard work in the brush leaves most men as pallid as corpses'.[7] Clement Hodgkinson, a private surveyor who worked in the northern New South Wales rainforests, observed the Macleay River cutters becoming subject to 'ague', a malarial fever, from the extensive surrounding swamps. He was also critical of the treatment of many, often ex-convict, cutters, who were dependent on the cedar dealers for payment at the end of a lengthy cutting season:

> *The cedar dealers furnish them from time to time with salt provisions, flour, tea and sugar; and every three or four months the sawyers travel down to the cedar dealers, who live at the mouths*

Cabbage tree forest, American Creek (NSW)
plate no 15, *Australian landscapes*, Melbourne, 1866–68
Eugene von Guerard
colour lithograph
Wollongong City Gallery
The George and Nerissa Johnson Memorial Bequest, purchased 1992

Von Guerard's lithograph records the devastation to the rainforest caused by the timber industry and those who cleared land for grazing with little regard for any subsequent effect on the environment.

Postcard
Cabbage trees, Illawarra, from the *Nature series*, c1910
photo lithograph, with printed colour
Private collection
With most of the rainforest cleared, it was often the lone, tall palms standing sentinel in a paddock that signified past botanical diversity and the once plentiful rainforests.

of the rivers, for a settlement of their accounts. As these latter individuals are not remarkable for delicate scruples of conscience, they generally settle the balance due to the sawyers in a very summary way. They take care to have a good assortment of clothing, tobacco, & c. in their huts, with which they furnish the sawyers at an advance of about three hundred percent on the Sydney prices; this, with a cask or so of rum and wine, to enable the sawyers to have a fortnight's drinking bout, generally balances their accounts. The scenes I have witnessed at the Macleay River, on these occasions surpass all description. Men and women, (for many of the sawyers have wives), lying day and night on the bare grass in a state of intoxication, and only recovering to renew their orgies; casks broken in, and the contents passed round in buckets; men fighting; native blacks, who have been supplied with liquor, yelling and screeching like demons, under the influence of alcohol.[8]

In the early days there were several reports of attacks by Aborigines. When the first sawyers reached the Nambucca River, the local tribe, which had been friendly to Clement Hodgkinson when he passed through in the late 1830s, attacked and killed several of the party. A great number of Aborigines were killed when a retaliatory expedition was raised.[9] As forests were often the last places of refuge for Aborigines, they were in close contact with the cutters and relationships ranged from friendly and helpful, to hostile.

Felling a huge tree required skill and hard work. Before any cutting, undergrowth needed to be cleared from the base of the tree. This often included the spiky 'wait-a-while' vine that could hook an unwary cutter. Judge Barron Field observed Illawarra cutters 'in the wild forest' where they had to perform the 'preparatory labour of clearing their path, and a fall for the trees, which would otherwise be prevented from reaching the ground by amazingly strong vines'.[10] The tree was then inspected and its lean and branch mass gauged to decide where it would be most easily felled.

Timber felling on the Richmond River, c1890
photographer unknown,
New South Wales
silver albumen photograph

Richmond River Historical Society Inc.

Two timber cutters use springboards to rise above the buttress of the tree.

The buttresses of rainforest species such as red cedar required the cutting team to set themselves up above the buttress in order to work freely with their axes and cross-cut saws. Sometimes a sapling platform was built, or steel-tipped planks notched into the trunk to serve as springboards. A man would balance on the board while wielding an axe or saw. An axe was used to cut the scarf, or V-shaped incision, which was made on the part of the tree facing the direction of the tree's fall. The bark was then peeled off the tree's stem right around from each end of the scarf. This stopped the teeth of the cross-cut saw, which was used for the cut back towards the scarf, from becoming clogged. Just in case the initial tree-felling judgment was incorrect and the tree leaned back and jammed the saw, wedges were hammered into the cut to hold the weight of the tree while sawing continued.

When the tree began to fall, a nimble evacuation of the site was necessary to avoid falling branches, the tangle of the underbrush, and falling debris from surrounding trees. If cutters were high above the buttresses on springboards, they had only seconds to get down to the ground safely. A tree might split before it was fully sawn through and the cutters could be thrown to the ground. Forest workers were particularly wary of the prophetic 'widow makers', dead branches which, without any warning, could break off high in a tree's crown and kill or maim anyone unfortunate enough to be standing below.

Once on the ground, the crown was cut off, branches trimmed, and the stem roughly squared and cut into lengths from 2.4 to 3.6 m for sawing onsite or for hauling to a timber mill for further processing. Timber processed onsite was pit-sawn, which involved levering or rolling the cut stem over a pit large enough for a man to climb into to work a cross-cut saw from underneath, while his lucky partner worked the saw from above. The stem sections were cut into boards or flitches, 25 to 50 mm thick, for easier removal from the forest and could undergo further hand-sawing or, as steam-driven saws came into use, mechanical sawing at a larger centre. According to Alexander Harris, the scent from the sawdust was so fragrant that a person working among it 'soon ceases to be sensible'.[11] Alternative sawpit designs included propping logs over an embankment and elevated timber platforms.

Two timber workers felling a tree on the Atherton Tableland, 1890–1900
photographer unknown, Queensland silver albumen photograph

John Oxley Library, State Library of Queensland (24043)

An axeman balances on a makeshift platform made from forest saplings while making the initial scarf cut.

Cedar getting on the Richmond River
after Julian Ashton, in Andrew Garran (ed), *Picturesque Atlas of Australasia*, published in London, in 1886
Research Library, Historic Houses Trust
Trees were cut into lengths from 2.4 to 3.6 metres for transport to sawmills by bullock teams.

Transporting the logs

Early in the 19th century, the task of transporting pit-sawn timber or log sections to their market destination was complicated and fraught. It involved travelling large distances, and being subjected to wet weather which could turn bush tracks into quagmires. When Barron Field toured the Illawarra in 1823 he observed:

> *The cedar planks, as they are formed by sawyers at the pit, are carried on men's backs up to the mountain summit, whence carts (approaching by a narrow road cut through the forest on the ridge) convey the planks to all parts of the colony, or they are carted to the shores of Illawarra, and navigated to Port Jackson in large open boats.*[12]

With improved transport in later years, logs, whole or in sections, were snigged by bullock teams to a collection point, or log dump. From here, one, two or three logs were loaded onto timber jinkers or pulled up on wagons drawn by teams of horses or bullocks for transport to sawmills. The 'bullocky' had to know his animals well – there were up to 20 in a team – to keep them hauling in unison. He was a deft hand with a whip that was most often used lightly and selectively, since, by all accounts, cursing was his stock in trade. Bullock teams and their handlers were integral to the timber industry until the early 20th century. In some instances logs were sent downhill to dumps on timber chutes, colloquially known as 'shoots', consisting of a groove gouged in the ground and lined with sapling logs. Like the early days, waste timber from each tree was tremendous, since only the best parts were transported to the mill.

If large, navigable waterways existed, logs were collected in the water, dogged together by chains or cables into rafts, and floated to their destination. At the journey's end, chains stretched across the river would collect the log rafts and millers would recognise their logs by brands that had been stamped into the ends. In cases where timber-cutters worked the mountain tributaries of rivers, they often chose to send the fallen logs singly down streams and rivers. Logs were also gathered on the banks of rivers to await river rises caused by heavy rains in the wet season. This was a wasteful method that left many logs stranded, waiting to become a lucky find for a new settler in later years. In the late 1880s, the North Queensland Timber Company sent thousands of logs over the Barron Falls for a 300-metre plunge before their journey towards the coastal port of Cairns. Not surprisingly, there was vast wastage as logs were smashed on the rocks below. Competition for logs among cutters was so intense that trees were sometimes felled without any chance of them being taken out of the forest – just to prevent other cutters from claiming them.

In places such as Kunghar Creek in northern New South Wales, contraptions known as 'flying foxes' were used to transport logs. These consisted of steel cables strung between firmly anchored points on the hillside and at the valley floor which carried the logs on rollers directly to the mill. Later in the 19th century and early in the 20th century, when the timber industry was on a more stable footing, logs from inaccessible areas were carried to sawmills by steam-powered engines that ran along specially constructed bush tramways. By the 1930s bush railways were superseded by roads, and caterpillar-tracked machines could venture into previously inaccessible places.

Milling the timber

Saw pits coexisted with sawmills throughout the 19th century. Huge pits, 15 x 7 m, complete with thatched roofs, were established in Moreton Bay in 1826. While much of the timber was processed in the forests, in the early days logs were transported to main centres such as Parramatta, Sydney and Moreton Bay for processing by hand saws. From the late 1830s, steam-driven mills were established in Sydney in 1838 and Brisbane in 1853. Capital cities and major towns remained the centres for timber processing long after nearby log supplies were exhausted. Later in the 19th century, steam-driven sawmills became common in medium-sized and larger towns and were often established throughout major timber-getting regions, in proximity to their raw material. Logs were stored in the yards before being processed into timber. Some mills only rough-sawed the green timber, which was sold to larger saw millers and merchants who air-dried it for greater stability and machine planed the wood. Generally, it was only large operators that had the capital and equipment necessary

Timber-getting, Canungra district, c1914
photographer unknown, Queensland
silver gelatin photograph
John Oxley Library, State Library of Queensland (64647)
In some of the most dense and inaccesible rainforest areas of New South Wales and Queensland, flying foxes, or aerial lines, were built to transport logs.

to process the timber for final use.

Steam power and improved saw blade technology revolutionised the industry and paved the way for the processing of huge volumes of timber. Nearly all of Brisbane's private housing and many public buildings were constructed of timber – the hoop pine, crows ash and red cedar from surrounding areas were almost completely exhausted before supplies from further afield were sought. In Sydney, where brick houses were standard, red cedar was used for window frames, skirting boards and other joinery, while the best quality timber was used for cabinetmaking. In both cities, major public buildings such as state parliament houses, courthouses, libraries, post offices and churches featured joinery of the best red cedar.

Impact on the Aboriginal people

It was inevitable that cedar-getters would come face-to-face with the original inhabitants of the forests. Reactions when the two groups met were mixed. Aborigines were often friendly towards the newcomers who were really only concerned with extracting the valuable forest timbers before moving on. Since most of the cutters applied the common principle of *terra nullius*, which assumed the land had no legal owners, they hardly paused to consider the consequences of their actions on an Aboriginal system of land tenure built up over the millennia. According to ethnographic and archaeological evidence, Aboriginal rainforest habitation would have varied from region to region. It appears that New South Wales and

Josias Hancock's saw-mills, Ipswich, c1888
Phillip-Stephan Photo-Litho and Typographic Process Co Ltd
Woolloomooloo, New South Wales 1884–1910
coloured photo-lithograph
Josef Lebovic Gallery

Huge timber mills continued to operate and process cedar well into the 20th century. They were seen as examples of efficient industry and Australia's primary wealth.

Natives spearing fish on the Bellinger River from Clement Hodgkinson, *Australia from Port Macquarie to Moreton Bay,* published in London in 1845

Rex Nan Kivell Collection, National Library of Australia

The rivers and forests provided Aboriginal people with regular food supplies which were greatly disturbed by forestry and land clearance for grazing.

southern Queensland Aborigines used the rainforests on a seasonal basis, which depended on both the climate and availability of food, while those in north Queensland lived more or less permanently in the tropical rainforests and developed a distinct culture based on their forest environment.

Clement Hodgkinson was greatly assisted by a tribe from the Bellingen district who enabled him to:

> *get rapidly through the entangled briars in the brush by beating down vegetation with their boomerangs.*[13]

Quite often it was Aboriginal men who climbed the trees to spot the new foliage which showed up the cedar trees, while in north Queensland there are reports of Aborigines working in timber-cutting operations.[14] Here also, it was often the Aboriginal pathways through the rainforest that facilitated 'exploration' of new areas. Carl Lumholtz, the Norwegian ethno-botanist, would not have found the rare tree kangaroo without the benefit of local Aboriginal knowledge, and explorer and journalist Archibald Meston believed that hundreds of lost settlers were saved by black trackers.

Once they established that the new arrivals were here to stay, Aborigines often decided that they needed to match the European's technology

Postcard
Clearing scrub lands, from the *Pioneer Life* series, c1906
Charles Kerry, photographer
photo lithograph, with printed colour
Private collection

Postcard
The Big Scrub, Richmond River, c1910
photo lithograph, with printed colour
Private collection

Postcards reflected general enthusiasm for images of progress.

by acquiring their tools of trade. Early cedar-getters in northern New South Wales, for instance, reported that the Aborigines 'were constantly watching them and would rush and seize their axes if they left them even a few yards away from them'.[15] When the Macleay River cedar-getters moved to the Nambucca district in the late 1830s, they were sometimes under siege. One report of the consequences of Aborigines killing and wounding several sawyers noted:

> *... retaliatory expeditions were undertaken, in consequence, against the natives, (on which occasion the sawyers mustered together, armed with their guns, and swords, roughly manufactured from their pit-saws) a great number of blacks were killed in the skirmishes which took place, and they gradually became more peaceably inclined.*[16]

Cedar-getters generally had a more benign influence on Aboriginal people than pastoralists who had their eyes on land for grazing stock. But while they may not have displaced the original inhabitants to such an extent, they still introduced diseases like chickenpox and new stimulants like alcohol. The cutters also pointed settlers in the direction of new potential farming land that eventually led to local Aboriginal people being displaced. By the time explorer Christie Palmerston was walking the ancient Aboriginal pathways through the rainforests of north Queensland late in the 19th century, attitudes to Aborigines had hardened: 'thin shields may answer very well for the purpose of their wars, but my rifle drilled them as if they were sheets of paper'.[17] Around the same time Archibald Meston, obviously influenced by Darwin's theories of evolution, reflected upon the Aboriginal situation in north Queensland as a 'mournful picture ... of the old inhabitants who for unknown ages have roamed the primeval forests moving off silent and swiftfooted into oblivion.'[18]

Impact on the environment

There were many accounts of the destruction of the rainforests – one observer summed up the situation in northern New South Wales in 1869: 'the devastating axe of the timber-getter has made dire havoc amongst the cedar brushes and where a few years ago immense quantities of wood were to be found there is not a single tree worth cutting'.[19] However, it was clearing land for agriculture that caused the greatest devastation. Logging for cedar was destructive

Farley's selection, c1900
James Ramsay
albumen silver photograph
Historic Photograph Collection, Macleay Museum, University of Sydney

By around 1900 maximum economic gain was considered best achieved by using every acre of a small holding. The potential environmental damage of these practices was not considered.

in that one species was almost obliterated, but this had little effect on the forests compared to the selection of land for farming. The rainforests would have recovered their cedar stocks through seedling germination and regeneration if settlers had not followed and cleared the forests for agriculture.

Clement Hodgkinson's initial reaction to the economic possibilities of the rainforests from Port Macquarie to Moreton Bay was that they had the potential to become plantations for sugar cane, cotton, coffee and rice, but he didn't believe they could be cleared unless cheap labour was introduced. When the Robertson land reform laws were introduced in New South Wales in 1862, land was opened for small settlers. Soon large areas of the north coast were suffering the same fate as the Illawarra, Shoalhaven, and Hunter river valleys earlier in the century. Around the turn of the century, photographers like Charles Kerry produced postcards that portrayed a romantic view of land clearance. Kerry's *Pioneer Life* series postcard titled *Clearing Scrub Lands*, is a sylvan scene of rainforests being cleared for settlement. By the 1890s most of the Richmond River forests and its Big Scrub rainforest had been felled, burned and ploughed to pasture. It proved easier than anyone ever imagined. A coloured postcard clearly shows the demarcation between a rainforest remnant and the pasture on which cows and horses contentedly graze. By the time the Farleys were clearing their selection in the Tweed around 1900, maximum economic gain was considered best achieved by using every acre of the small holdings. When the tree cover on the steep hills was removed, heavy summer rains washed the topsoil into the waterways. Before long, navigable rivers were nothing but a series of shallow pools. And, ironically, silted river mouths spelled the end for river transport.

By 1910, about half of the coastal rainforests of south-eastern Australia had been cleared for agriculture. In some cases it took only one generation for the pristine forests to become farmland. One writer summed up the change to the environment by the 'great invading wave of white faces':

> *The old things are fast passing away, the black faces have become fewer and are dying out and ... the gorgeous birds have hidden themselves in the thickest scrub. It is not all gain to see the peace and silence of nature chased away by the axe of the settler and the noisy activity of the steam engine.*[20]

With most of the rainforests cleared, it was often the lone, tall palms standing sentinel in a paddock that signified past botanical diversity and the once plentiful rainforests. By the turn of the century, many city dwellers had formed a romantic attachment to the ferny dells of the mountain forests outside capital cities such as Sydney and Brisbane, where journeys were made easier by the establishment of new rail services. They were taking day trips to emerging tourist spots such as the Blue Mountains at the time when areas like the Big Scrub were being destroyed.

Red cedars in plantations

As early as the 1880s trial plantings of cedars were attempted in New South Wales, both in mass and as individual trees. There have also been trial plantings in various Queensland state forests. But attempts to establish red cedar plantations have been thwarted by the cedar tip moth (*Hypsipyla robusta*), which lays its eggs in the tips of growing leaves. On hatching, the larvae tunnel through the growing shoots of young trees, causing the terminal shoot to die and the tree to branch. The effect of many moths in the tips of a cedar is for the tree to heavily branch at its crown. Since timber production calls for long, straight stems, the tree becomes unsuitable for commercial use. When red cedars grow as single specimens within rainforests, they are not as often attacked, or if they are, the trees are tall enough for the timber potential not to be so seriously affected.

The CSIRO has used seeds collected from

Postcard
Mount Tambourine, Queensland, from the *Queensland Views* series, c1910 photo lithograph, with printed colour
Private collection
Postcards of popular scenic spots captured the pristine natural qualities sought by people who were becoming increasingly aware of the beauty of the Australian bush.

Felling a cedar tree, Gloucester, NSW, and ***A cedar tree, Gloucester, NSW***
1944
Fred Barrie
photographic dye, gelatin silver photographs in cedar frames

Newcastle Regional Museum

Bought for £900 in 1944, the wood from this tree was used by Fred Barrie, a Newcastle cabinetmaker, for special furniture mostly made for himself and his family.

various places to establish field trials in Australia and several Pacific Island countries to assess genetic variation in growth characteristics and tip moth resistance. They hope to select genetically resistant strains, improve management practices and, eventually, re-establish red cedar as a commercially farmed tree in Australia and South Pacific countries. These days mature red cedars in the wild are rarely seen and usually confined to inaccessible gullies or national parks and reserves that were gazetted many years ago.

By the end of the 19th century, the red cedar forests of eastern Australia were largely logged out. The quest for this beautiful and valuable tree that economically sustained two fledgling colonies was the direct cause of great human suffering for many thousands of displaced Aboriginal people who found themselves in the path of the cedar-getters. It also led to the widespread devastation of vast tracts of subtropical and tropical rainforest from southern New South Wales to north Queensland. These excesses and the unsustainable manner in which the species was exploited underlie the story of extreme human effort that went into harvesting the 'red gold' and working with the excellent timber. Today the rich legacy of red cedar survives in the form of finely crafted furniture, objects and building joinery – a tribute to Australia's early carpenters and craft workers.

3

'... LIKE A LONDON GIN PALACE!' RED CEDAR IN THE AUSTRALIAN INTERIOR

CLIVE LUCAS

The stranger is much struck by the handsome appearance given by the profuse use of cedar in the fittings of the Sydney dwellings. It has all the beauty in colour and figure of the Spanish mahogany; indeed, the experience of an upholsterer is necessary to detect the difference by sight alone. In solidity and closeness of grain the Australian cedar is, however, greatly inferior to mahogany. The doors and sashes, the window frames and shutters, staircases and balustrades, skirting boards and cornices, and, in a few instances, the floors and ceilings, are of cedar. Even the housemaids' closets have all the exterior appearance of polished mahogany doors. This profusion of dark-coloured unpainted wood in the fitments of a house pleased my eye exceedingly; but my taste was disputed by many – some going so far as to assert that it made a dwelling-house look like a London gin-palace!

LIEUTENANT-COLONEL GODFREY MUNDY WRITING IN 1848.[1]

The large quantity of unpainted cedar used in skirting boards, staircases, doors, window frames, architraves, reveal panels and window sashes in mid19th-century Australian houses was an unusual sight for English eyes. The English equivalent of cedar, mahogany, was used sparingly and really only for door leaves, which were often veneered, and stair handrails. Its use was mostly restricted to upper-class buildings, with lesser mortals having to make do with painted graining, in imitation of mahogany.

English 19th-century houses, notably those in the gothic revival style, used oak for the total fit-out. This was the closest the English interior came to resembling the Australian early Victorian wooded interior.

There is much confusion in Australia between Georgian and early Victorian architecture. A lot of what has traditionally been regarded as Georgian is in fact early Victorian, with the interiors more influenced by the Greek style neo-classicism following the Napoleonic Wars than by the feminine style of the Adam brothers, fashionable Scottish architect designers of the late 18th century Georgian period. Characteristic of the early

Dining room, *Cressbrook*
1842, Queensland
Courtesy the McConnel Family
Photograph Ray Joyce
Several early Queensland houses have interiors of unpainted cedar. Entering them is like climbing into a large cedar wardrobe and closing the door.

Victorian interior was the disappearance of the surbase and the introduction of high skirting boards. The column no longer sat on a pedestal, but came firmly down to the floor.

All of the so-called 'colonial Georgian' buildings such as Camden Park (1831), Panshanger (1831), Elizabeth Bay House (1835), Clarendon (1838) and Fernhill (1840), are, in fact neo-classical early Victorian buildings in the Greek revival style with the exception of Clarendon which is in the French neo-classical style.

Box pews, St Thomas's Church
Port Macquarie, New South Wales
Royal Australian Historical Society
In districts where cedar was forested and milled, abundant use was made of the wood in interiors.

By early Victorian times there was a taste in Australia for the fully polished interior of the type Mundy commented upon and this has a lot to do with neo-classicism and the masculine robust nature of the Greek and gothic revivals. However before 1830 Georgian interiors were often painted and/or grained as they were in England and the joinery timber was not always cedar, but often, hardwood or pine.

In 1816, for example, Francis Greenway, the first Colonial Architect of New South Wales was specifying New Zealand pine for the Rectory house at Parramatta. Presumably this was kauri pine and he suggested that it be painted. The following is taken from his specification for the Parramatta Parsonage, dated 15 April 1816.

> *If the doors and windows shutters be of New Zealand Pine, with Architraves, Backs, Elbows, and Soffits, Base and surbase they must be painted, if cedar they may be left without paint ...*[2]

Early colonial architecture was popularised in the 1960s by the National Trust with a particular focus in New South Wales and Tasmania on the major pre-1850 buildings which were all generally in the neo-classical style with polished joinery. This led to a frenzy of stripping because of a perceived belief that all colonial interiors had polished cedar joinery. As a result, many colonial buildings got polished joinery for the first time and, in some cases, had the joinery replaced when it was found to be of different woods or too difficult to strip. This was the case in the 1960s restoration of Old Government House at Parramatta. Here the joinery has recently been repainted in lead paint in its late Georgian colours.

Generally, panelled work, doors, reveal shutters and embrasures were made of cedar. This was as much to do with its properties as a good, workable joinery timber as with the character of the wood. Naturalist and surveyor Clement Hodgkinson wrote in 1845, 'the cedar is employed in Sydney for every purpose to which deal is generally applied; and is also used in all kinds of cabinetwork, as it is of a handsome grain and colour'.[3]

The few pre-1830 buildings that survive intact feature mostly painted joinery, whether

Drawing room, Old Government House
1799
Parramatta,
New South Wales

National Trust of Australia (NSW)

Photograph
Christopher Shain

Painted, not polished, joinery, often made of a mixture of cedar and other woods, was popular in the early 19th century.

cedar, hardwood or pine. Door leaves and window sashes, however, like stair handrails, were often left unpainted, not so much for look as practicality. Greasy hands were better on polished work than on lead based paint that could not be cleaned.

From the time cedar was discovered it was used for joinery, often in conjunction with pine when it was available. Pine was first imported from New Zealand and later from the Baltic region, and later still, North America. Cedar was the only native timber that was suitable for joinery which did not blunt joiners' tools.

From the late 1820s neo-classical taste made cedar the chosen joinery timber for appearance. It was lavishly used in the fit-out of all the buildings constructed after this time until the depletion of the cedar trees.

In Sydney the abundant use of the wood can be seen in buildings such as Elizabeth Bay House and Lindesay, and New South Wales churches such as the Pitt Street Congregational Church, Sydney (1841), St Thomas's, Port Macquarie (1824), and St John's, Stroud (1833).

Cedar was also widely used in the interiors of Tasmanian buildings, even though it didn't grow there. While the joinery of window frames and sashes were often pine, Tasmanian houses like Panshanger (1831), Ballochmyle (1835), Woolmers Cottage (1838), Morningside (1839) and Killymoon (1842); churches like St

Legislative Council Chamber, Parliament House, from 1856, Melbourne

Courtesy Parliament of Victoria

Government buildings often made copious use of cedar.

Doorway, *Aberglasslyn*
1840, Maitland, New South Wales

Photograph Max Dupain, 1978

Late 1820s neo-classical taste made cedar the chosen joinery timber for its rich effect. It was lavishly used until the depletion of the cedar trees.

Synagogue
1843, Hobart

Photograph Frank Bolt, 1979

Queen Victoria Museum and Art Gallery, Launceston

The contrast between painted, unpainted and gilded surfaces creates a rich effect in the interior in Australia's oldest synagogue.

George's, Battery Point (1836) and St Andrew's, Evandale (1840); and other public buildings like the Legislative Council Chamber (1856) and the Hobart Synagogue (1843) make copious use of cedar in their fit-outs. Other neo-classical houses in Tasmania such as Highfield (1826), Bicton (1837) and Clarendon (1838) featured pine joinery which was originally painted or perhaps grained.

Cedar seems to be an eastern seaboard phenomenon. In South Australia one finds more graining employed and in the far west of the country, early buildings used the local hardwood jarrah for fit-outs of the 'gin-palace' variety. A noted example is Minniup (1848) south of Bunbury where the joinery is distinctively bold, only making use of beads because of the problem of running mouldings in such a hardwood. Often in Western Australia the unpainted feature of an early Victorian interior was restricted to a chimneypiece, usually of jarrah. For example, the main staircase at Government House in Western Australia is polished jarrah while in Sydney it is cedar. Both houses are gothic revival in style.

Of all the places it has been used, the taste for unpainted cedar reached its apogee during the colonial period in Queensland. Here not only the doors, windows, skirting boards and chimneypieces, but the lining boards to the walls and ceilings and sometimes even the floors were made from cedar. Entering an early Queensland house such as Cressbrook (1842) or Coochin Coochin (1843), is like climbing into a large cedar wardrobe and closing the door. Everything is unpainted cedar and the resulting interiors are extraordinary. In other examples such as Mundoolun (1850s), the external face of the walls was of unpainted vertical sawn cedar slabs.

Mundy's mention of cedar floors may have been in reference to Queensland (in 18[illegible]8 it was

Reading Room, Fisher Library
1920s

Rare Books Collection University of Sydney

Photograph Edward Searle

Designed by the New South Wales Government Architect Walter Liberty Vernon, the Fisher Library was perhaps the last great use of cedar in Australia.

still part of New South Wales) for there are no known cedar floors in New South Wales today, except for stair landings.

Other than the building itself, there were often examples where interiors were complemented by purpose-built cedar furniture as in the fitted library bookcases at Camden Park and at Elizabeth Bay House (now sadly destroyed). Many houses had fitted cupboards, sideboards or bookcases beside chimney breasts. Notable examples are in the painted room at Rosedale (1828), the library at Lake House (1830) in Tasmania and in the dining room at Lanyon (1859) in the Australian Capital Territory.

As the century progressed, houses tended to use less cedar. While a terrace house of 1850 would have been completely fitted out in cedar, by the 1880s probably only the staircase would have been of cedar and sometimes only the handrail. Increasingly, cedar was saved for mansions and important public buildings, in particular courthouses and churches. This use was, however, generally confined to New South Wales and Queensland where the cedar tree grew. Victoria and the other states tended to use pine and graining. A major country house like The Gums (1876) in Victoria features a cedar staircase, while at Burrungurroolong (1880) in New South Wales, the whole fit-out is cedar.

That is not to say that cedar was always left for polishing. Particularly in the feminine rooms such as drawing rooms the woodwork was painted and this was the case at Burrungurroolong.

The high point for the use of cedar was probably the 1870s after which its use declined as the cedar trees were logged to virtual extinction from the rainforests of the eastern seaboard.

By the 1890s conservative New South Wales country houses like Anambah (1890) and Boree Cabonne (1897) were still fully fitted in cedar but houses that didn't want painted schemes, and buildings that were inspired by the teachings of William Morris and the Arts and Crafts movement often chose kauri pine from New Zealand and other Australian woods such as Queensland maple and hoop pine. Alton (1882) in Victoria, Rona (1893) in New South Wales and Harrow (1893) in Queensland, are major examples of this aesthetic taste.

While in England a quality gothic revival church of the Victorian period would have been fitted out in oak, in Australia the pews, the altar, pulpit and so on were all of cedar. It was only as the supply of cedar diminished that oak or kauri pine fittings started to appear in the antipodean church. It is interesting to compare two Hunter Valley churches designed by the architect Edmund Blacket. St Mary's at West Maitland dates from the 1860s and is fitted out in cedar whereas St Peter's at East Maitland dating from the 1880s is all in kauri pine.

The buildings which lavishly used cedar right up until the end of the century were major courthouses, council chambers and public buildings. The General Post Office in Sydney (1874), the Law Courts in Melbourne and its library (1881), the Adelaide Town Hall Council Chamber (1882), the Supreme Court in Bendigo (1892) and the Banco Court in Sydney (1895) are *tour de forces* of the joiner's art, as are the Victorian Houses of Parliament (from 1856) and in particular, the Legislative Council Chamber and its spectacular library (1860).

Interior of the Rare Books and Manuscripts Reading Room
National Library of Australia, 1967–69, Canberra
National Library of Australia
Photograph David Reid
With matching furniture by Fred Ward, the walls are lined with cedar panels not unlike some of the earliest colonial interiors.

The old Fisher Library at the University of Sydney (now alas minus its bookcases), designed by the New South Wales Government Architect Walter Liberty Vernon and built in 1902–09, is perhaps the last example of the rich use of cedar in Australia. Its interior is memorable for its Art Nouveau detailing and fittings. There is another library, the Rare Books and Manuscripts Reading Room at the National Library of Australia, which is lined with cedar with chairs and tables ensuite. Designed by the Sydney architects Bunning & Madden and built in 1967–69, this is perhaps the very last gasp.

Cedar had become a luxury item to be used only on very special occasions.

4

FURNISHING A COLONIAL PALACE: GOVERNMENT HOUSE, SYDNEY

ANN TOY

> *I have at last got into the New House. We dined and slept in it for the first time on Thursday ... The moving has been a work of no small labour.*[1]

Considering the length of time from consent to completion – 13 years in all – taken for Sydney's new Government House to evolve on the edge of Farm Cove, this announcement of occupancy sent in June 1845 to Sir George Gipps's colleague Charles La Trobe was highly restrained. Gipps had good cause for frustration. Having been appointed governor in 1838, his entire incumbency up until the move had been spent in anticipation of a building initially approved by the British Home Office in 1832 but not begun until 1837. Sadly, he spent just over a year in his new residence, leaving office in July 1846 and dying the following February in England.

The governor was also disconcerted by the economics of the project. The edifice had not only taken more than seven years to construct but its final cost of £46,077 represented a budget over-run of £21,000. In addition, he was having to deal with the economic depression in New South Wales and the mounting insolvencies of the colonial gentry. Surprisingly, however, little criticism was directed at the scheme's extravagance. General opinion and the press regarded the new Government House as a symbol of the colony's advancement.

There is no question the building was of paramount significance for New South Wales. The imperative to both relocate and create a residence in keeping with the regalia and symbolism of the vice-regal office had been implemented only after decades of protest by successive governors about the inadequacies of the existing Government House (at the corner of Bridge and Phillip Streets). Believing no architect in the colony capable of designing such an important structure, Governor Bourke requested that the Colonial Office engage 'an eminent architect' in London. The commission was duly awarded to Edward Blore, 'Special Architect' to William IV. By late 1834 a full set of working drawings for a Tudor gothic revival

Main staircase

Government House
Historic Houses Trust

Photograph Brenton McGeachie

Detail of the grand cedar staircase and gothic arched screen in the main hall, designed in 1834 by Edward Blore. The carved gothic lamp supports were supplied by Andrew Lenehan in 1857.

Royal arms of the United Kingdom
c1867
James Cunningham
cedar, paint and gilding

Government House
Historic Houses Trust

Photograph Brenton McGeachie

The work bears the incised signature of the carver, 'J Cunningham'.

residence was dispatched to Sydney.[2] The design was a clear statement of a specific intention to maintain 'in so remote a dependency of the Empire some of the visible state and splendour which belongs to Her Majesty's Representative'.[3]

Since its establishment in 1832, the Office of the Colonial Architect was principally responsible for the design, construction and maintenance of public buildings, including the decoration and furnishing of interiors. Appointed Colonial Architect in 1835, Mortimer Lewis was to supervise the construction of the new Government House and was given discretionary power in the selection of building materials. In accordance with Blore's plans the building was constructed of sandstone. For the interiors, detailed specifications were provided for the plasterwork, painted finishes, joinery and structural features. The designs for the cedar door and window joinery, wall panelling and skirting were relatively simple. But the rich-toned cedar magnificence we see today in the halls might have been lost to the world had it not been for Lewis's intervention. Ignoring Blore's direction to grain the joinery 'in imitation of oak', Lewis chose instead to have the cedar varnished to enhance its natural colour.[4] This choice of local materials such as red cedar, native marbles and sandstone was deliberate, largely influenced by factors such as their availability, cost and their comparable quality with imported timbers and marbles. Their use was a major technical and stylistic achievement as this had never been attempted on such a grand architectural scale. The new Government House had a seminal influence in raising the standards of building and craftsmanship in Australia.

Throughout the history of the acquisition of furniture for Government House, Sydney, cedar has played a major role. Though not complete, available documentation of government expenditure on maintenance, re-decoration and refurbishment confirms an extraordinary and important collection of original cedar furniture dating from the mid 1840s to 1900, much of which is still in use today.

Predictably, the records also reveal that the choice of furniture was subject to the vicissitudes of fashion and the changing tastes of governors and their wives, often influenced by the Colonial Architect. As furniture fell out of fashion, it was gradually relocated upstairs to the private apartments or into storage. Over time much of it was auctioned off, transferred to other public buildings, placed into government stores or discarded.

Governors were expected to pay a substantial amount towards the cost of furnishing and decorating the private apartments while at the same time keeping a tight control over expenditure for the public rooms. One imagines this was a burden on those without adequate

Government House Sydney
1848
George Edwards Peacock
oil on board

Government House
Historic Houses Trust

Photograph Ross Heathcote

private means. From an annual salary of £5000 Governor Gipps, for instance, was required to pay staff salaries, underwrite entertaining costs and foot the expense of keeping and providing household effects for two official residences.[5]

Ever resourceful, Gipps transferred the best and most serviceable furniture and furnishings from the first Government House and Government House, Parramatta. Next, pointing out 'the size and number of the public rooms', he wrote to Lord Russell requesting an unusually large sum of £3000 to help furnish 'the New House'. This request was approved.[6]

Presumably it was this augmentation to the annual furnishing budget that enabled Mortimer Lewis, in February 1845, to purchase the principal contents of the drawing room of Elizabeth Bay House, the residence of the former Colonial Secretary Alexander Macleay. Most of the rosewood furniture and accoutrements had come from London in the late 1830s, but they were evidently fashionable enough for Lewis to contemplate re-using for the vice-regal drawing rooms.

One of the Colonial Architect's duties was to compile an inventory of the public furniture at Government House Sydney and at Parramatta at the end of each governor's term of office. Lewis made the earliest inventory for the new Government House in 1846. A comparative study of this and later inventories, combined with private and official correspondence, pictorial records and the physical examination of original furniture, has verified the importance and wide-ranging use of cedar furnishings in both of these residences. In 1846 the drawing rooms featured four cedar card tables, a round table, twelve chairs and three easy chairs, despite a taste by this time for imported English or European furniture. Rosewood or mahogany, in particular, were considered de rigueur among the 'pure Merinos' and the nouveaux riches in the colony.

In conformity with this prevailing British taste, a quantity of new rosewood drawing-room furniture was also commissioned from Irish-born craftsman Andrew Lenehan, described in 1851 by Governor Sir Charles Fitzroy as 'an ingenuous and reputable cabinetmaker'.[7] The original commission consisted of a pair of Brazilian rosewood veneer sofa or library tables, a smaller matching table, a loo table, a pair of card tables and armchairs in the rococo revival style. Most of this furniture is still in use today and importantly, cedar was used for the carcass, or framework, of these pieces. An insight into the competitive world of cabinetmaking is provided by a vigorous dispute between Lenehan and a previously unknown cabinetmaker, John Arbuckle, which was documented in *The Sydney Morning Herald*, 19 March 1846, following the *Herald*'s attribution to 'Mr Lenehan of Castlereagh Street' of some 'massive and carved work ... of considerable beauty' furnished by him for the new Government

Drawing room

Mitchell Library, State Library of New South Wales

In 1873 the drawing room was refurbished by the firm David Jones. This photograph shows some of the original Lenehan rosewood furniture: a twin pedestal sofa table and a card table still in use. The large suite of maple drawing furniture is believed to have been supplied by John Hill in the 1860s.

Sofa table
c1845
Attributed to
Andrew Lenehan
Brazilian rosewood,
cedar (secondary)
Government House
Historic Houses Trust
Photograph
Brenton McGeachie

House. Arbuckle's letter to the paper accused Lenehan of vanity, claiming that 'every particle of massive and carved work at present to be seen in new Government House' had been performed by himself. He specifically claimed to have made the 'massive throne chairs'. Lenehan responded with a list of his own, enumerating specific chairs and tables 'together with various other articles both decorative and useful, too numerous to mention, with which I have been favoured by His Excellency'.[8]

If we can believe Arbuckle's claim it would appear that the four 'Elizabethan style' chairs now displayed in the main hall are his handiwork. These cedar chairs and an accompanying hall table – no longer in the collection but depicted in a Conrad Martens sketch of the hall circa 1855– are based on designs in J C Loudon's *An encyclopaedia of cottage, farm and villa architecture and furniture …*[9] The appropriateness of furnishing a vice-regal residence with furniture intended for a modest villa may seem questionable, but their large scale complements the interior.[10]

Surviving accounts, official correspondence, inventories, newspaper articles and Lenehan-provenanced furniture still at Government House all help to establish the fact that his firm received continuous and prominent vice-regal patronage from 1845 to the late 1860s. Like some of his fellow cabinetmakers such as Charles N Hunt, Henry Woolley and John Hill, Lenehan's business provided a variety of services ranging from the supply of furniture, curtains and upholstery, floor coverings, light fittings, pictures and framing, to the repair and refurbishment of old or damaged furniture.[11]

Vice-regal patronage of local cabinetmakers was a longstanding tradition, as most of the early domestic furniture requirements were initially met first by marine, and then by convict carpenters and cabinetmakers. Early governors also recognised the value of trade in native timbers, particularly cedar, and encouraged its use in official buildings. This patronage was principally motivated by economic considerations: the lower cost of

Hall in Government House, Sydney, September 10th
1855
Conrad Martens
pencil drawing

Dixson Galleries, State Library of New South Wales

Martens prepared this sketch with detailed notations of the wall colours, textiles and furniture. He recorded the cedar staircase, wainscotting and the early furniture, including three of Arbuckle's hall chairs.

colonial furniture and the need to support the growth of local industries and employment.

The *Sydney Morning Herald* 11 March 1846 was prompt to acknowledge Governor Gipps's support:

> *It is pleasing to note that an example has been set in high quarters which will have the effect of discouraging the purchase of foreign or even British furniture for our own woods and our own mechanics can now supply and manufacture articles of beauty superior even to those brought to us across the water.*

Inventories of government-owned furniture at Government House, Sydney and Parramatta compiled in 1821 record over 80 pieces of locally made furniture using cedar, rosewood, beefwood and blue gum. At Parramatta, their placement throughout the principal rooms of the house suggests that many were of some quality.[12] Again, the inventories of 1831 show that a large quantity of cedar furniture was still being used alongside the English mahogany furniture imported by Governor Darling.[13]

The contracts for the masonry and carpentry work associated with the new building were also awarded to local firms, much of the labour sourced from the growing pool of newly arrived immigrant artisans attracted by the colony's favourable immigration policy. The recent discovery of a Return of 'The number and description of Mechanics employed in the Department of the Colonial Architect Sydney', dated 13 March 1847, and sent by Lewis to the Colonial Secretary, throws new light on the original furnishing of Government House. The document lists four cabinetmakers and carpenters: Edward Ledger, Samuel Jefferies, Charles Snow and James Snow engaged on the work of 'Making furniture [for] New Government House' at the rate of 3/6 per annum. A polisher, James Gaul, worked 'cleaning furniture for new Government House' for 2/6 per annum.[14]

We do not know precisely the type of furniture these men made for Government House. Circumstantial evidence suggests that in the capacity of 'estate carpenters' they possibly made and installed cedar and/or pine 'back of house' furniture such as built-in cupboards, shelves, presses, dressers, kitchen tables, food safes, benches, stools, serving trays and trestle stands.

Later in the 1860s, additional cedar furniture was supplied by other local cabinetmakers. John Hill, for example, supplied a quantity of cedar office furniture and carried out extensive repairs and modifications to the Lenehan dining suite;[15] G C Hayes, described as 'one of the best workmen in the colony', made a large cedar press for the Housekeeper's room and Henry T Jones made two presses, bookcases, a large double pedestal desk and a pair of stools. These pieces survive in the collection.

The end of Sir John Young's term as governor (1861–1867) coincided with a major refurbishment of the State rooms in anticipation of a visit in 1868 by Prince Alfred, the Duke of Edinburgh. It was the first royal visit in Australia's history. In preparation, tenders were called in 1866 for '6 Ottomans or Seats for Ball room in all 55 feet 6 ins long covered with best silk Taboreatte (sic) at the cost of £97.12.6.'.[16] It is believed that Lenehan was awarded this contract.[17] The ballroom is currently furnished with eight cedar ottomans or banquettes. Those featuring cabriole-shaped front legs were supplied (between 1866 and 1872) by the firm Charles Artlett & Son, earning it a royal warrant. The other banquettes were not supplied until later, between 1871 and 1879. They feature turned front legs, and were the work of Alexander W Norton.

During this period two cedar royal arms of the United Kingdom c1867 were supplied by woodcarver James Cunningham, similar pieces were made for Hartley Court House (1863), the Supreme Court, Sydney (1870s?) and Kempsey Court House (1897). Cunningham's work was sought after for both architectural and ecclesiastical commissions.[18]

After Lenehan's retirement in 1875, Alexander W Norton's Hunter Street business became one of the principal furniture contractors to Government House. Among his earliest vice-regal commissions were four stylish gothic hall chairs in cedar, now displayed in the inner and outer halls. With their elaborately carved pinnacles, quatrefoil decoration and vertical pierced backs the chairs are similar to patterns published by Loudon and in Thomas King's *The cabinet maker's sketch book of plain and useful designs* (vol 1, 1835).

Linen room, ground floor of the service wing

Government House
Historic Houses Trust

Photograph
Brenton McGeachie

Several of the cedar presses are marked with an 1872 inventory stamp but their maker cannot be identified.

The ballroom
c1887

Government Printing Office Collection, Mitchell Library, State Library of New South Wales

This view records the interior with the ottoman seating prior to its extension by Government Architect W L Vernon in 1899.

opposite
Outer hall with Alexander Norton Hall chairs
c 1870
cedar

Government House
Historic Houses Trust

Photograph Brenton McGeachie

Originally a tall vaulted porch, the outer hall was remodelled in 1873 as an internal vestibule and furnished with pairs of hall chairs by A W Norton. The cedar coats of arms commemorate the occupancy of the house by the Governors-General of Australia from 1901–12

Other existing examples of Norton's work at Government House include two cedar bookcases, a wardrobe and several pine toilet mirrors. There are also display cabinets and a pair of tables with decorative veneers in native timbers such as Huon pine, tulipwood and acacia. From the mid-1870s, when these lighter timbers and walnut became fashionable in drawing rooms, the popularity of cedar began to decline. Over the next two decades the resurgence of interest in Chippendale, Adam and Sheraton revival-style furniture led to an increase once more in the popularity of rosewood.[19]

In 1894 Government House was the beneficiary of a major international event, the World's Columbian Exposition of 1893 held in Chicago, USA. The occasion was significant because New South Wales was the only Australian colony to be represented. The New South Wales Commissioners, with the encouragement of Sir Henry Parkes and Sir George Dibbs, had resolved to create a contribution that could pose 'before the world as the embodiment of Australasia'.[20]

The New South Wales Government spent a record amount of £29,928 in order to showcase a comprehensive display of the colony's staples of wool, minerals and timber as well as promote its potential in agriculture, fisheries, horticulture, transportation and viticulture. The centrepiece was an exhibition of colonial manufactured furniture intended to provide the major impact of the Forestry Department's promotion of commercial native timbers.[21]

Consisting of three suites for library, bed and dining rooms, this furniture was shipped home to Sydney's principal residence. The Australian entry had failed to win any major Exposition awards, possibly because its design and

The Right Honourable Sir JOHN ADRIAN LOUIS HOPE
Earl of Hopetoun P.C. K.G. G.C.M.G. G.C.V.O. First Governor
General of the Commonwealth of Australia 1901-1902
The Right Honourable Sir William Humble Ward
Earl of Dudley P.C. G.C.M.G. G.C.V.O. Governor
General of the Commonwealth of Australia from 1908-1911
The Right Honourable
General of the Commonwealth of Australia
The Right Honourable HALLAM
Baron Tennyson K.C.M.G. Governor
the Commonwealth of Australia 1902-1903

craftsmanship were conventional in comparison with the visionary examples of Art Nouveau furniture exhibited by the French.[22] Yet in the development of decorative arts in the colony, the library suite made by Verdich & Co, of Darlington and the bedroom and dining suites by George Bartholomew & Co, York Street, Sydney, are outstanding late 19th-century examples of Australian art furniture.[23]

Uniformity of decoration had been ensured by a Forestry commissioning committee anxious not only to 'illustrate the suitable character of some of our principal timbers for the purpose of the cabinetmaker' but to incorporate emblematic Australian designs.[24] One of those commissioned to carve the decoration on the Bartholomew exhibits was Queensland-born sculptor Harold Parker. *Art in Australia* in 1922 claimed that he was only 17 at the time.[25] It was a direction to be developed after 1901 by craftsmen such as Robert Prenzel into a distinctive Australian form of design.[26]

Companion pieces in the Verdich display included three occasional tables of colonial timbers, a hallstand of colonial rosewood and a blackbean doorway. Bartholomew's dining-room furniture, comprising a sideboard, buffet, table and 12 chairs, was also of blackbean timber, and the suite of bedroom furniture – wardrobe, dressing table and wash stand – of local beech, silky oak and purple wood. Both these suites were also decorated with carvings of Australian flora, New Zealand ferns and Maori-inspired spiral motifs. *The Building and Engineering Journal* of 11 June 1892 claimed the designs for the Bartholomew furniture were 'from the pencil of Mr William A Ker', artist and designer. Given the similarity of the motifs, it is likely Ker was also responsible for designing the Verdich suite.'[27]

Upper drawing room or boudoir
1900

Beauchamp collection, Mitchell Library, State Library of New South Wales

Photograph showing the Verdich secretaire bookcase and marquetry inlaid rosewood furniture by local cabinetmaker C N Lykke – who was formerly in partnership with Verdich.

Secretaire bookcase
c1892
Edward Wilhelm Verdich
cedar, colonial rosewood

Government House
Historic Houses Trust

Photograph by
Brenton McGeachie

The existence of the Verdich suite – comprising a secretaire bookcase, an octagonal shape table and six chairs – is important as no other examples of furniture provenanced to his company have been located. The height (266.5 cm) and scale of the inverted breakfront secretaire bookcase is impressive. It was clearly intended to be a tour de force, designed to showcase the utility and beauty of red cedar. Featuring carved panels of Australian flora and distinctive friezes of Maori-influenced curvilinear spiral designs, technically the carving is bold and vigorous. Verdich's expertise is also evident in the finesse of the repoussé copperwork escutcheons and handles.[28]

The synthesis of New South Wales native timbers, Australian flora and vegetation motifs, traditional Maori decoration and New Zealand flora is intriguing.[29] The concept behind the commissioning of this furniture must be examined in light of contemporary discussions about Australia's role in supporting British Imperial control of the South Pacific and the growing national and international political debates, involving countries like Fiji and New Zealand, around Federation. The iconography chosen for the New South Wales Court at the 1893 Exposition was therefore required to be symbolic, projecting a sense of unity between the respective British colonies within the broader regional context of Australasia. It also needed to reinforce the rise in nationalistic sentiment following the centenary celebrations of European settlement of Australia in 1888.

The last colonial governor of New South Wales was William Lygon, the seventh Earl Beauchamp (1899–1901). A cultured man, Beauchamp had an impressive private art collection and a passion for the English Arts and Crafts movement, a taste he shared with the Government Architect, Walter Liberty Vernon.[30] Together, their selection of new office furniture struck a different sensibility for Government House.

Characterised by uniformity and modern design, it stood in direct contrast to the randomly assembled collection of cedar office furniture sourced over the years from different cabinetmakers. The Arts and Crafts gothic revival style furniture by Walker Sons & Bartholomew Ltd were distinguished by the lighter and more mellow hues of oak, their simply carved panels, flowering capitals of stylised foliate decoration and elegant Art Nouveau metal fittings. It is important to be aware, however, that as with some of the rosewood furniture, the oak is merely a veneer. Its cedar carcass indicates it was made in Sydney.[31]

Following Federation, from 1901 to 1912, Government House became the Sydney residence of the Governor-General of Australia

Dining room

Government House
Historic Houses Trust

Photograph
Brenton McGeachie

Andrew Lenehan's most outstanding vice-regal commision was the principal suite of furniture c1857 for the dining room. The table and sideboard's designs are derived from the English furniture pattern book, The cabinet maker's assistant *published by Blackie and Sons, London 1853.*

The study
Government House
Historic Houses Trust

Photograph
Brenton McGeachie

View of the study featuring part of the extensive suite of Arts and Crafts gothic revival style oak office furniture, c1899, commissioned from the firm of Walker, Sons & Bartholomew, Sydney.

under an arrangement between the New South Wales Government and the new Federal Government. During this period the State governors resided at the house Cranbrook in Bellevue Hill. According to the lease documents and contemporary inventories, the furniture collection remained in situ. Alternatively, the new vice-regal residence and Hillview, the governors' country residence at Moss Vale, were furnished by local manufacturers: Beard Watson Ltd, Farmer & Co, Anthony Hordern & Sons, David Jones and James Lawson & Sons.

In the early 1980s Government House was again transformed by a major restoration of the historic interiors by the Department of Public Works under the supervision of conservation architect Donald Ellsmore. Research and re-cataloguing of the furniture collection undertaken in conjunction with historian Kevin Fahy led to the discovery and restoration of many original cedar items in storage at Government House and the Department's depot at Sheas Creek.[32] In 1999 the Historic Houses Trust of New South Wales introduced the policy *To Furnish a Future*, to re-establish an innovative link between contemporary Australian design and craft and the refurbishment of the State Rooms of Government House.

The author wishes to acknowledge the assistance of the following people in the preparation of this essay: Beverley Earnshaw for her research contribution, Dr Rosemary Annable, Rick Bartholomew, Roger Blackley, Dr James Broadbent, Dr Jenny Blain, Donald Ellsmore, Kevin Fahy AM, Penny Gill, Ray Gurney, Ross Heathcote, Joy Hughes, Jane Kelso, Elaine van Kempen, Terence Lane, Robert Lowrie (Parliamentary Archivist, Parliament House New South Wales), Dr Judith McKay, Roslyn Maguire, Megan Martin, John McPhee, Jessie Serle, Edgar Verdich and Keith Verdich.

5

'OF THE BEST CEDAR PROCURABLE ...' CEDAR IN SYDNEY TOWN HALL

MARGARET BETTERIDGE

Sydney Town Hall, the city's pre-eminent expression of civic authority, was constructed in two major stages between 1875 and 1889 using nothing but the finest the colony could provide in stone, timber, metal and glass. For the architects, designers and craftsmen engaged on the project, it was an opportunity to showcase their talent on a grand and very public scale. Red cedar felled from the north coast of New South Wales provided a particularly rich and handsome vehicle for the carpenters, joiners and French polishers to demonstrate their prowess. Their craftsmanship in Sydney Town Hall is today considered to be unique and among the finest known local joinery and carving to survive intact in its original setting.

Cedar was significant during the five major phases of decoration in Sydney Town Hall, namely the Ante Chamber or Vestibule (1876–1878); Ground Floor Corridors and Offices (1878–1880); Council Chamber (1883–1884); Centennial Hall (1886–89) and the Lady Mayoress's Rooms (1905–06). It was used exclusively for all of the decorative joinery embellished to reflect the influence of the stylistic trend contemporary with each phase. With the exception of the furniture for the Main Hall and The Lady Mayoress's Rooms, the major items of furniture for public rooms and offices were made of cedar.

Although the town of Sydney was formally incorporated as a city in 1842, it was to be two decades before the city council secured their preferred site for municipal administration. The colonial government, keen to keep the council within its sights in the government precinct, repeatedly offered them the site of the first Government House for which James Barnet, later Colonial Architect, proposed a 'grand desideratum' and offices in the Italian style.[1] The aldermen, however, pursued a relentless campaign for the Old Sydney Burial Ground, located deep in the heart of their power base in Sydney's fast growing commercial and retail area, and threw open the design of the town hall and municipal offices to public

Town Hall, Sydney
c1880
William Cooper
pencil, pastel and wash
Sydney Town Hall collection, City of Sydney

The design for Sydney Town Hall was a fusion of fashionable French Second Empire style and the traditional classicism of contemporary English and American town halls.

competition. The winning entry by J H Willson, subsequently amended by city architect Albert Bond and city engineer Edward Bell, was a fusion of fashionable French Second Empire style and traditional classicism, no doubt inspired by contemporary town hall architecture in England and America. The story of its design and construction is a long and sorry one, punctuated by episodes of cost over-runs, professional misconduct, fraud, poor workmanship and character assassination. The haphazard nature of the surviving records reflects the turmoil behind the scenes and much of the story, including the use of cedar in the town hall, is incomplete.

The Ante Chamber (Vestibule)

The first instruction for the use of cedar for Sydney's new town hall was included in the tender documents of 20 July 1875 for the completion of the Ante Chamber (Vestibule). Cedar 'from slow grown large trees carefully cut'[2] was specified for the ceiling spandrels around the dome, the curved ribs of the ceiling and panels to support the lathing, the coving pieces of the soffits, the window framing for the double-hung upper storey sash windows and the sill of the dome light. In addition to its use as structural and framing timber, the carpenter and joiner were 'to fix carved coupled cedar scrolls with bed moulds and swags on the face of each upper pedestal, and to supply the main entrance doors, 3-inch thick folding with enriched stop column, to be bolection moulded[3] carving and cut externally, moulded raised panels internally ...'.[4] Apart from the new sashes and frames which were to be coated on one side with linseed oil, all wood work was to have three good coats of white lead oil paint and two flatting coats to match the plaster. The

Pair of doors
from an alderman's office, Sydney Town Hall, 1878
David McBeath, designer
red cedar
Sydney Town Hall collection, City of Sydney
Photographs Brenton McGeachie
Henry Smith and James Bennett's execution of the design for doors in the Sydney Town Hall resulted in some of Australia's most elaborately carved joinery.

Interior design for Sydney Town Hall, 1878
David McBeath, designer
pen and ink
City of Sydney Archives

City architect McBeath's designs included elaborate cedar joinery to complement other aspects of the Italian Renaissance style of decoration.

main external doors were finished with two further coats of colour and three coats of approved varnish enriched with gold bronzing.

Ground Floor Corridors and Offices

Henry Smith and James Bennett of Burrahpore Street, Woolloomooloo, who undertook the work, were the successful tenderers in November 1878, engaged to complete the lobby, entrance, corridors and rooms on the ground floor which enclosed the Ante Chamber. Their contract for the carpentry and joinery specified that 'the cedar for the joinery will be of the best quality, it must be well seasoned and free from shakes, knots, cracks or other defects'[5] and that it be finished to receive French polish. Hand-coloured elevations drawn by David McBeath, then city architect, provided a detailed visual reference for the cedar joinery, including drawings of the carving on the doors and mouldings to complement the elaborate plasterwork, marble fireplaces, gilded overmantel mirrors and decorative light fittings.[6] Italian Renaissance was the predominant theme for the enrichment of the carved joinery, with floral lozenges and lion masks on delicately carved scrolling foliage door panels framed within classically inspired torus and ovolo mouldings, rosette and palmette bands and stele crests. While many of the designs can be traced to pattern book ornament, it is the sophistication with which the wood has been carved that suggests the influence of imported skills. The name of Ranieri Paolicchi has long been associated with the carved cedar in Sydney Town Hall, but records which authenticate his work have yet to be located. Although many of McBeath's

Perspective sketch of the Council Chamber from *The Building and Engineering Journal*, September 1888 attributed to John Hennessey

Mitchell Library, State Library of New South Wales

The Council Chamber's joinery with complementary furniture was described as being of 'very choice cedar' and arranged in such a way as to be 'in a high degree original and effective'.

designs were sourced from pattern books of the time, some of the finer details were likely to have been discussed and resolved on site.

The carved lion masks and dolphins on the front door and the laurel wreath enclosing the Wiesner clock at the entrance to the original Mayor's Office on the ground floor, may have been designed with symbolic intent but there is nothing recorded to prove it. The early call in the specifications to the plasterers for 'Australian flora and fauna [to be] introduced wherever the City Architect may require even when figures may not be previously described or shown in the enlarged details'[7] did not extend to carved cedar decoration.

The Council Chamber

With offices on the ground floor completed for the mayor, the aldermen and staff, a meeting room for council became the next priority. The design of the room and its suite of furniture has been attributed to John Hennessey who was employed by the municipal council in 1881 as a draughtsman and later, until his resignation in 1884, as assistant city architect to city architect Thomas Sapsford. Hennessey had received his early architectural training at London's Royal Academy of Arts where he gained practical experience with Queen Anne revival architect Basil Champneys, artist Charles Eastlake, and noted gothic revival architect, William Burges.

An architectural perspective of the chamber was published in *The Building and Engineering Journal* in September 1888 with an accompanying description of the room which made particular reference to the joinery and furniture. '... The woodwork, fittings being of a very choice cedar, French polished the rich colour of the wood falls in well with the grey brown and red of the curtain draperies ... The whole room is in such excellent taste that it would be difficult to particularise, but may we pay direct attention to the unique design of the arrangement and detail of the Aldermanic benches and of the mayor's chair, the latter especially being in a high degree original and effective.'[8]

Ground floor corridor
Sydney Town Hall
City of Sydney
Photograph Greg Piper
Classically inspired motifs are used to decorate the elaborate joinery and plasterwork of lobby, entrance, corridors and ground floor rooms.

Hennessey was undoubtedly influenced by the teachings and writings of Eastlake, who espoused aestheticism for the simplicity and honesty of its design, favouring the furniture styles of early English Jacobean, and medieval German. Elements of these are reflected in furniture crafted for the Council Chamber, particularly the panelled backs to the upholstered benches, the bulbous feet to the mayor's chair and aldermanic benches, the turned cross- framed stretchers to the large meeting tables, and the carved Tudor rose paterae, acorns, oak leaves and strapwork embellishments, to which were added quarter-circle fluted fans, a signature motif of the Aesthetic movement.

The contract for the Council Chamber, including the manufacture of the furniture, was awarded to William Coleman, builder and contractor, whose business address was 59 Sydney Arcade, City. The value of the work was £10,975. Work commenced in February 1883, and the room was ready for occupation the following year. For over 120 years, the Council Chamber has been the stage for city governance and the platform for vigorous debate. On several ceremonial occasions during the 20th century, the complete suite of furniture was removed from the room and re-assembled in the Vestibule and the Main Hall.

In 1972 the furniture arrangement in the Council Chamber was reversed to provide better viewing of council proceedings from the first floor gallery. During the dismantling of the canopy and screen behind the Lord Mayor's chair, workmen discovered the following notation in pencil on one of the timber panels '... Adam Brooks born April 10th, 1862 at Leith, Scotland fitted up this canopy and chair February 15th 1884'.

View of the Council Chamber
Sydney Town Hall
City of Sydney
Photograph Eric Sierins
William Coleman built the imposing mayoral canopy to the design of the assistant city architect. Like the other furniture of the room, it favours English Jacobean and medieval German styles.

View of the Council Chamber
Sydney Town Hall, 1909

City of Sydney Archives

Hennessey's designs for furniture in the Council Chamber are influenced by fashionable English theory and combine elements of historical styles with Aesthetic movement motifs.

Coleman subsequently submitted a large claim for extras totalling almost £2000, which included lining the staircase to the public gallery above the room and alterations to the bench ends and the large councillor's table, the supply of '24 chairs with carved and turned work, polished and upholstered in real morocco' at a cost of £11. 8s each, an additional 'Councillor's table with 12 drawers, crystal knobs, covered with cloth and polished' (£31. 12s) and 'a table for reporters with carved and turned legs, covered with cloth' (£8. 12s).[9] This claim for work additional to the original contract and requested by Thomas Sapsford, prompted an enquiry by council's finance sub-committee, who determined Sapsford to have been negligent in his management of the contract. Coleman was finally awarded his costs in March 1886 after lengthy deliberations by council, who considered his claims to have been excessive.

The Main Hall

The second major building stage of the construction of the town hall was well underway when Sapsford, his career in tatters, died suddenly in late 1886 at the age of 39, leaving the completion of his great hall to his successor, George McRae. Again, 'cedar, the best procurable, seasonably cut and as old as can be obtained'[10] was specified in the building contract for all joinery work. The external doors to Druitt Street and the Cathedral were to be '3-inch framed and moulded with raised mouldings and panels both sides with carved paterae and mouldings'[11] to designs supplied by the city architect. John Hennessey, who had worked with Sapsford on the original designs, was re-engaged by council as a consultant in 1887 and returned to supervise construction of the hall. He is thought to have been responsible for the detail, including the design of the wood panelling apron enclosing the stage, and decorative motifs such as the carved lyres on the newell posts on the stage stairs. The care with which the timber had been selected and fashioned attracted praise from many reviewers. *The Builder and Contractors' News* of 7 December 1889 noted that 'the woodwork of the galleries is all of the very best selected cedar' and *The Morning News* of 15 November 1889 wrote that 'the galleries are of the best cedar, and with their dark polish form a pretty contrast to the white walls and lighter timber flooring'.

Lady Mayoress's Rooms

Cedar, polished to glow a rich, warm, red-brown colour throughout the town hall, was used to its greatest dramatic effect in the decoration of the Lady Mayoress's Rooms. This suite of three first floor rooms on the northern side of the building was created in 1890 in recognition of the increasingly important role of the Lord Mayor's wife. In 1905, Council authorised the removal of the walls between the rooms and the construction of large openings in place of doors, at an estimated cost of £100. Mountney and Company of Pyrmont supplied the rolled steel joists and girders to support the masonry, and council employees completed the cladding of the openings, plastering and

Architectural drawing of door openings, Lady Mayoress's Rooms, 1905
H Merriman
pen and ink
City of Sydney Archives

The assistant city architect's design for these door openings combine classical elements with the fashionable Art Nouveau of the early 20th century.

opposite

Door frames, Lady Mayoress's Rooms
City of Sydney
Photograph Greg Piper

The bold and imposing cedar door surrounds helped to create a suite of Parisienne style reception rooms.

re-painting. Drawings survive for bold soaring upswept cedar surrounds in Art Nouveau style prepared by assistant city architect J H Merriman, who created a suite of Parisienne style reception rooms which have been the backdrop for official entertaining ever since. In accordance with the finance committee's instruction, the work was undertaken by carpenters and joiners employed by council.[12]

Mayoral chair
from the Council Chamber, Sydney Town Hall, 1883
attributed to John Hennessey, designer
William Coleman, manufacturer
red cedar, leather
Sydney Town Hall collection, City of Sydney
Photograph Greg Piper
The large mayor's chair, designed to be seen against the imposing mayoral canopy, continued the theme of combining historical and aesthetic elements.

Furniture

Cedar was prominent in the hierarchy which dictated the furnishing of rooms in the town hall. While imported Thonet-style bentwood chairs provided the public seating in the Main Hall and the Lady Mayoress's guests sat on dainty reproductions of Sheraton and Hepplewhite sofas supplied by Beard Watson & Company, the working furniture was more robust. The offices of the Mayor and Town Clerk were each furnished with large partner's desks, a large armchair with carved top rail and leather upholstery, and a glass fronted bookcase, all in cedar. Desks, drawing tables and plan cabinets furnished the offices of the city engineers, surveyors, architects, sometimes designed and drawn by staff. Bookcases, large filing presses and writing tables were provided for clerical staff, the quality of the timber always a high priority. William Coleman included a typical suite of furniture in his claim for additional payment in 1885. He supplied the water department with four of his best cedar office tables with drawers, movable desks and paper presses with pigeon holes and sliding doors; a cedar press with six doors, turned work and brass locks and a cedar counter with return end flap door, made according to plans at a cost of £148.[13] His claim also included a cedar framed booklift for the Treasury Department. In keeping with Sydney Town Hall's curatorial policy, recent contemporary furniture has been commissioned in cedar, but its lack of availability has meant that alternatives have been sourced overseas.

Sadly, almost all the furniture is unmarked save for a Council inventory stamp. It would seem that much of it was manufactured by contractors, or purchased from suppliers, rather than made in-house. Maintenance and repair of the cedar joinery and furniture, however, fell to the team of staff carpenters, joiners and polishers who, for many decades, operated from workshops located in the basement of Sydney Town Hall, many completing a four year apprenticeship. By the close of the 20th century the last of the in-house carpenters and polishers had retired or re-located, casualties of modular furniture, synthetic materials and rising labour costs. Shortly before their departure, the town

Main Hall
Sydney Town Hall
City of Sydney
Photograph Greg Piper
When completed the Main Hall was praised, 'the galleries are of the best cedar, and with their dark polish form a pretty contrast to the white walls ...'

hall revealed one last cedar secret. Buried below them were the remains of some of the earliest known cedar to be used in the colony. Fragments of timber coffins from the Old Sydney Burial Ground[14] were uncovered during archaeological excavations outside their workshops and forensic timber analysis identified the wood, although badly decomposed, as having the characteristics of Australian red cedar. Not only then does Sydney Town Hall have some of Sydney's finest cedar joinery – it is also keeper of some of its earliest.

The author wishes to acknowledge the assistance of Dr [illegible] Murray, Research Historian, City of Sydney, Stephen T[illegible] Manager, Properties, City of Sydney and Lois Sabine in the preparation of this article.

6

CARPENTERS, CABINETMAKERS, WOODCARVERS AND CRAFT WORKERS

JOHN MCPHEE

The origins of furniture making in Australia evolved from necessity and basic practices. On arrival the military personnel and convicts from the First Fleet sought local materials for building essential furniture such as beds, chairs, benches and tables to supplement the meagre supply they brought with them. Some of the earliest furniture made in the colony would have been beds made from saplings and branches lashed together to make legs, and a frame to support a 'mattress' of rushes cut from the edges of the swampy harbourside bays – their simple construction followed an ancient tradition. Similarly, the first improvised chairs and stools would have been equally simple lengths of log to sit on. Benches and roughly constructed tables would have been the next items of necessary furniture, and they too would have been cut and assembled from green timber close at hand.

Today the needs of campers, itinerant workers and pioneer settlers anywhere in the world remain similar. Later Australian settlements associated with the gold fields continued the pattern. Writing about life on the gold fields in the middle of the 19th century traveller G B Earp described this simple furniture: 'Chairs grow abundantly in the New South Wales forests as any moderately sized tree cut into two feet lengths will testify.'[1] However no documented examples of the earliest furniture survive and with the unfortunately typical Australian desire to forget both convict and humble origins, it is unlikely that there would have been any desire to preserve them.

The timber that was immediately available for the colony's earliest makeshift furniture would have been the casuarinas and eucalypts of Botany Bay and Port Jackson.[2] More suitable timbers were found as the surrounding countryside was explored. In March 1791 William Bradley, the first lieutenant on the *Sirius* and a diarist, identified 12 trees from which he thought the timber showed promise for furniture making, among them a 'bastardised mahogany' which was probably

Bowl
1986
Grant Vaughan
red cedar
Collection Powerhouse Museum, Sydney
Inspired by the sheath of the fruit of the Monstera deliciosa this bowl is a rare example of contemporary Australian craftwork making use of cedar.

Australian red cedar.[3] With an eye to the possibility of economic benefit, samples of various timbers were regularly sent back to England. In 1795 a cargo of a timber described as an Australian mahogany and cedar was sent back to England.[4] While it is not known what happened to this specific cargo, English botanists, timber merchants and cabinetmakers would have assessed similar consignments. Writing in 1803, the London furniture designer Thomas Sheraton suggested that the colony 'remains in general still uncultivated, uncivilised and in a great measure unexplored',[5] inferring that timbers of economic use were yet to be found. However he described the four samples of timber he knew for certain to be of Australian origin as being of 'hardish texture, and some of it very ponderous'.[6] In English society, where most furniture was made from West Indian mahogany, and crisp carved decorative detail was fashionable, the difficult-to-work and unpredictable exotic woods of the colony were not favoured. At best, they were deemed suitable only for inlays or for small items, such as tea caddies, which became souvenirs and curiosities of the distant colony.

In the colony, carpenters and cabinetmakers made use of Australian cedar for furniture soon after its discovery and by the 1820s, if not earlier, it was in regular use, not only in New South Wales but also in Tasmania where no similar easily worked dark timber was available. Compared with the tough and unpredictable eucalypts and other woods of the colony, cedar was easily cured and not difficult to work. It did not blunt the poor quality tools available to most of the earliest colonial furniture makers. The dark reddish colour resembled mahogany and took a high polish. This helped further liken cedar's appearance to fashionable English mahogany and explains its almost universal popularity for locally

Framed timber specimens, c1850, in the hallway, *Camden Park*, Menangle, New South Wales, from R T Baker, *Cabinet timbers of Australia*, 1913
Private collection
These timber specimens, or duplicates, are thought to have been exhibited at the Exposition Universelle *held in Paris in 1855.*

Side chair, c1820
maker unknown, Tasmania
red cedar, replacement upholstery
Private collection
Photograph Brenton McGeachie
This simple chair was made by a relatively unsophisticated chair maker with some knowledge of fashionable English style or to meet the demands of a more knowledgeable client.

made furniture for most of the 19th century.

The earliest furniture makers in the colony would have been either attached to the military, ships' carpenters, or were among the few convicts with carpentry and cabinetmaking skills. Working with unfamiliar woods, poor tools, and in most instances, only a familiarity with provincial styles, they made furniture to meet the demands of the colony. The first buildings requiring furniture were houses for the governor and military officials, government offices and buildings, and later, the few free settlers who had prospered enough to need furniture for their newly built houses and, in some instances, colonial mansions. At first, the supply of imported furniture was meagre, but recent research suggests that more furniture was imported from India than previously imagined.[7] This would have supplemented the furniture that local merchants and speculators imported from Britain as well as the production from the government lumberyards and workshops and a few local cabinetmakers.

On estates, assigned convicts and free men with carpentry skills were often responsible for making the furniture that filled a growing household's requirements. While utility was usually the most important quality of this furniture, and the makers had little or no knowledge of anything beyond mere function,

Settle, c1830
maker unknown,
New South Wales
red cedar, hardwood,
replacement
upholstery
National Gallery of Australia

Some early 19th-century furniture found in Australia looks remarkably Indian and may have originated there or been inspired by Indian designs more suited to a hot climate.

Sofa, c1840
maker unknown, Tasmania
red cedar, brass castors, replacement horsehair upholstery
National Gallery of Australia

Towards the middle of the 19th century the neo-classical motifs employed by cabinetmakers were often enlarged and exaggerated.

many examples bear witness to a desire to emulate European fashions. In some instances those for whom the furniture was made, had ideas about style, from either a recollection of furniture fashionable when they were last in England or from the few pattern books available in the colony. Some, most probably those with experience with the military in India, sought furniture that took into account the antipodean climate. Open-backed settles and chairs with caned seats, often with loose palliasses, are examples of the desire for less upholstered furniture that accommodated the climate, as well as being less formal in appearance.

Little cedar furniture exists from the earliest days of the colony which can be reliably dated or associated with a known interior. One of the earliest examples is the D-end dining table made in about 1820 for Richard Rouse's new house, Rouse Hill, at Parramatta. Rouse's responsibilities as Superintendent of Public Works at Parramatta included the government Lumber Yard, and it is reasonable to consider that the table was made there. Simply made, the table with two D-ends and a central section with drop flaps, has square tapering legs and follows a type fashionable in furniture pattern books published in the late 18th century before turned legs superseded the earlier style.[8]

Undoubtedly the most fashionable and sophisticated example of cedar furniture documented as having been made in a government timber yard is a desk made at the King's Yard,[9] Hobart, in 1835. Made from high quality and beautifully figured cedar imported from New South Wales, this unusual desk, which follows no known pattern, is believed to have been made for Governor Arthur's use at Government House. An unusual form, perhaps commissioned to meet the user's particular requirements, the secretaire cupboard rests on a tabletop with drawers and is supported by a lyre pedestal base of considerable elegance. The desk is marked with the names of its convict makers George Wilkin and Robert Graham,[10] the former being a house carpenter transported to Tasmania in 1829. The desk is an excellent example of the highest quality of cabinetmaking in cedar that occurred in the colony in the early 19th century. A few cabinetmakers worked to produce furniture to match their memory of fashionable English furniture and in rare instances they were able to resort to up-to-date pattern books. The most remarkable example of this practice is a pair of bookcases which were inspired by a design published in G Smith's *A collection of designs for household furniture and interior decoration*, 1808, London. Made by Joseph Sly soon after his arrival in the colony,[11] the pair of bookcases is a spectacular example of colonial cabinetmaking.[12] While inspired by Smith's design, the bookcases omit the wings, secretaire drawers and astragal bars of the original design and there is some variation of decorative features. However the biggest divergence from the original design must be the use of a dark timber instead of the lighter coloured oak that Smith indicates preferable and which was most fashionable for gothic style furniture in the early 19th century. The softness of cedar and its inability to carry carved detail also gives the finished bookcases a less crisp effect.

Another example of colonial furniture following a published pattern is the 'gothic' style chairs made in about 1845 for the entrance hall of the newly built Government

Desk, 1835
George Wilkin and Robert Graham
red cedar, other Tasmanian timbers
National Trust of Australia (Tasmania)
Photograph Andrew Simpson, courtesy Casuarina Press
This desk was made by convict craftsmen in the government timber yard in Hobart.

***D-end of dining table,** from Rouse Hill House, c1820*
maker unknown, New South Wales
red cedar
The Hamilton Rouse Hill Trust Collection
Photograph Ross Heathcote
The D-end dining table from Rouse Hill was probably made in the government Lumber Yard at Parramatta. Its simple construction is a reflection of the limited tools and skills of its maker.

Design for a bookcase
from George Smith's *A collection of designs for household furniture and interior decoration*, published in London in 1808
Photograph courtesy Casuarina Press

Bookcase, (one of a pair)
c1840
Joseph Sly
cedar, glass
The Australiana Fund
Photograph Andrew Simpson, courtesy Casuarina Press
Sly's bookcases omit some features of Smith's design and are made of dark cedar rather than light-coloured oak.

House and which are copied, almost faithfully, from J C Loudon's *An encyclopaedia of cottage, farm and villa architecture and furniture ...* First published in 1833, Loudon's eminently practical book was soon available in the colony. In it he suggests that this' ... parlour chair could be turned in oak, chestnut or almost any other British wood that has colour and variety in its texture' and that the '... design would be a cheap one'.[13] The carpenter/cabinetmaker, probably John Arbuckle, who made the four chairs worked in cedar, not following Loudon's suggestion for the use of a lighter coloured wood.

Early 19th-century cabinetmakers delighted in the abundance of cedar, the ease with which it could be worked, and, at times, its impressive grain. Occasionally, especially for tilt-top centre or loo tables, where the surface could be featured when the table was not in use, spectacular panels of cedar were used to dramatic effect. The symmetrical, or book-matched, panels of highly figured cedar, especially kept by the cabinetmaker for such effects, often made for a remarkable display of the timber's grain.

While not as dense or hard as mahogany, the favoured timber of woodcarvers, cedar was easily carved, although its softness did not permit the fine detail and crispness of European cabinetmaking. Some carvers specialised in architectural decoration such as the royal coat-

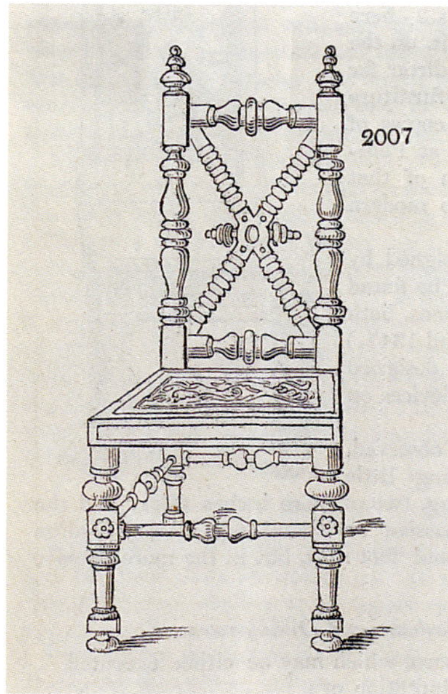

Design for a parlour chair from J C Loudon, *An encyclopaedia of cottage, farm and village architecture and furniture ...* published in London in 1833
Private collection

Hall chair, 1846
from *Government House*, Sydney
attributed to John Arbuckle
red cedar
Government House
Historic Houses Trust
Loudon's design for a parlour chair was adapted and made larger for the hall of Sydney's most fashionable gothic styled Government House.

of-arms for the Supreme Court, Sydney, and carved furniture. Of these, the best known is James Cunningham[14] whose work in the Sydney Town Hall shows the quality and nature of craftsmanship fashionable in the late 19th century. His lectern presented to All Saints Anglican Cathedral, Bathurst, in about 1872, is one of the most impressive examples of late 19th-century Australian carving in cedar.

Cedar was exported to other settlements and used by cabinetmakers as well as builders in Tasmania, Victoria, and South Australia, where cedar did not grow. These trades made use of the plentiful supply from New South Wales, and, later in the 19th century, from Queensland. In South Australia the German immigrant cabinetmakers made furniture to suit the needs of their clientele, other immigrants familiar with a style popular in their homeland.[15] In South Australia's Barossa Valley, in particular, but also in other areas settled by German immigrants in western Victoria and around Brisbane, furniture was made reminiscent of the European Biedermeier style. While much of the furniture made by the South Australian cabinetmakers was in pine, some painted and grained to look like the traditional timbers of German cabinetmaking, a few examples were made from the rarer and more expensive New South Wales cedar. In Victoria and Queensland, cedar, imported and local, was used and more commonly polished rather than painted.

The gold rushes created a large and increasingly affluent society in need of furniture to equip their houses. The most plentiful and easily worked local timber that was suitably dark coloured was, of course, cedar and huge amounts of very ordinary furniture were made throughout the 19th century. Not only houses, but government buildings, ate up the supply. Banks and post offices were lavishly fitted out in cedar joinery and panelling and had furniture to match. More impressive buildings had carefully designed furniture to complement sometimes lavish interiors. The furniture for the Sydney Town Hall chambers and joinery such

Lectern, c1873
attributed to
James Cunningham
red cedar

All Saints Anglican
Cathedral, Bathurst

Photograph
Graham Lupp

Cabinetmakers and woodcarvers enjoyed working with cedar because it was easy to carve and its rich colour and grain looked good when polished.

opposite

Centre table, c1840
Unknown, Tasmania
red cedar, rosewood

Private collection

Photograph Andrew Simpson, courtesy Casuarina Press

Cabinetmakers delighted in opportunities to display the grain of carefully selected and 'book matched' or 'butterflied' cedar.

as featured doors, written about in Margaret Betteridge's essay, were typical of the best use of the cabinetmaker and carver's skills in late 19th-century furniture.

The cabinetmaker and carpenter's tradition of furniture making in Australia was complemented by the less sophisticated activities of homeowners 'making-do' and building their own furniture. While there are many examples of simple cottage furniture made from cedar by carpenters meeting local needs, there are surprisingly few examples of the kind of furniture described as 'primitive'.[16] It would appear that cedar, although easily worked, and in the forests where it was being cut, readily available, was found to be unsuitable for everyday furniture in a settler's hut. The relatively soft cedar was prone to splitting if not carefully and strongly crafted so it was likely that other more hardy timbers more suited to simple construction methods were used in preference. Of course, the failure of examples to survive may not mean that it was not made, but that it just did not stand up to the rigour of everyday usage and has therefore been thrown away.

Late in the 19th century when a centenary of European settlement was celebrated in 1888, a nationalist movement gathered momentum and writers such as Henry Lawson and the painters of the Heidelberg School began to celebrate things peculiarly Australian. Artist craft workers made use of native and local timbers in their furniture and frequently decorated this furniture with carved details of Australian flora and fauna. The most famous of these cabinetmakers was Robert Prenzel[17] whose career in Melbourne was particularly illustrious. He favoured blackwood (*Acacia melanoxylon*) and blackbean (*Castanospermum australe*) in particular throughout his career, probably because they were hard timbers, not unlike oak or mahogany, with which he would have been familiar from his training in Europe. He helped make these and other Australian native timbers popular with those people interested in modern furniture.

Sofa, 1840s
maker unknown, Geelong, Victoria
red cedar, replacement upholstery
Gift of Michael and Traudl Moon, 1996
National Gallery of Victoria
Immigrant cabinetmakers used Australian timbers to create furniture in familiar styles, in this instance imitating the German Biedermeier

In Sydney, although designers and teachers such as Lucien Henry[18] and R T Baker[19] passionately promoted Australian motifs in the decorative arts and Australian timbers, there appears to have been no significant cabinetmaker to take up their challenge. Most turn-of-the-century cabinetmaking in New South Wales employing Australian timbers and with Australian floral and faunal motifs is the work of amateur craftworkers or one-off exhibition pieces. However, a significant, although aesthetically confused, group of works by E W Verdich[20] survives to demonstrate that there were some examples of nationalist furniture produced in Sydney. Ann Toy's account of furniture at Government House, Sydney, considers the origin and importance of this suite.

The growth of a leisured middle-class in Australia and the development of craft work of various kinds as a fashionable hobby, or a genteel way in which to supplement an income, led to the production of numerous objects made by craftworkers in the late 19th and early 20th centuries. While there are examples of amateur and professional craft work from early in the 19th century which make some use of cedar, it is not until later in the century that more interesting examples exist making greater use of cedar.

Wood carving had become a popular craft for men, and more adventurous women,[21] and cabinetmakers, such as Robert Prenzel, had numerous students. Others learnt their craft from magazine and newspaper articles or at classes held in the newly developed schools of art.

The fashion for woodcarving, produced numerous pieces of furniture with discretely nationalist and stylised art nouveau motifs, including screens and blanket and wood boxes decorated with gum-leaves and occasionally kookaburras. Chip-carving and fret-work were similarly popular crafts. Cedar was a popular wood for these crafts, presumably because it was easy to work and a common source, cigar boxes, was plentiful. Cigar boxes which were usually made of West Indian cedar, but occasionally from Australian cedar,[22] were frequently decorated with chip-carved decoration. While most examples are rather amateur in execution, some exist showing a more sophisticated sense of design and better than average carving skills. Occasionally a more ambitious object might combine chip-carving with pyrography, or poker-work, as the practice of burning a decoration onto a piece of wood with a hot wire, even occasionally making use of the sun and a magnifying glass, was known.

Some chip-carving, using panels of cedar from cigar boxes and pine from packing cases, was layered and carved to create objects we now describe using a late 19th-century American term, 'tramp-art'.[23] While the manufacture in Australia of these objects, usually small boxes and picture frames, was the domain of the amateur craftworker rather than the itinerant making objects for sale, they certainly belong to a craft tradition of little

A room in Woroongundi: a squatter taking it easy c1847
attributed to Charles Archer
watercolour and gouache

John Oxley Library, State Library of Queensland (67030)

In the districts where cedar was found houses were often made entirely from the wood, including floors, wall and ceiling linings. Bush carpenters also made simple furniture such as the bed and table shown here.

Frame, with chip-carved decoration, c1900
T Porschoest, Australia
red cedar
Private collection

Box, with chip-carved decoration, early 20th century
maker unknown, Australia
red cedar
Private collection

Cigar box, with chip-carved decoration, early 20th century
maker unknown, Australia
unidentified timber
Private collection
Photographs Jenni Carter

sophistication and less financial cost. Occasionally, craftworkers developed a more professional attitude to their work, and in South Australia, there developed among the German immigrant population of the Barossa Valley, in particular, a fashion for small objects, mottoes and even larger pieces of furniture, featuring fret-work, in which cedar panels were contrasted with lighter coloured wood such as pine.

Although, as noted by Baker in 1913, cedar fell from favour as a preferred cabinetmaking timber in the 20th century,[24] it is a cabinetmaker from a later wave of immigrants who again makes use of peculiarly Australian woods. Schulim Krimper arrived in Melbourne from a troubled Europe in 1939 and established a studio-workshop making furniture for a select clientele committed to contemporary furniture for modernist interiors.[25] He made extensive use of native Australian timbers, in particular blackbean (*Castanospermum australe*), but also jarrah (*Eucalyptus marginata*) and New Guinea walnut (*Dracontomelum mangiferum*), while also using the more fashionable teak. Like Prenzel, Krimper obviously sought out the harder timbers in preference to cedar but in several pieces he made special use of plain-grained timber to contrast with lighter timber for detail and decorative effect.

The most significant use of cedar in cabinetmaking in the later 20th century is in the Rare Books and Manuscripts Reading Room of the National Library of Australia, Canberra. Bunning and Madden's Library, completed in 1969, included a reading room based on the idea of a place set apart to read special books. For the room, panelled with cedar, Fred Ward

Chest of drawers on stand, c1955
Schulim Krimper
red cedar, Queensland silver ash
Art Gallery of South Australia
Ken Myer Fund 1992
Photograph Clayton Glen

Fleeing war-torn Europe Schulim Krimper celebrated his new life in Australia by making use of unfamiliar and exotic timbers from Australia and the tropics.

opposite right
Puzzle frame, c1900
maker unknown, Portland, Victoria
red cedar, gelatin silver photograph, glass, tin
Private collection
Photograph Jenni Carter

designed complementary tables and chairs.[26] They reflect the need for utilitarian and contemporary furniture for a building that was an uncompromising example of its period's sense of what was modern.

The taste for paler panelling and cabinet timbers in the later 20th century, possibly combined with the scarcity of suitable cedar meant that little was used in the Mitchell, Giurgula and Thorpe Parliament House, Canberra, of 1988. Some cedar veneer was used in the marquetry panels depicting Australian flora, nardoo (*Marsilea drummondii*) and *Idiospermum australensis*, designed by Tony Bishop and made by Michael Retter in the entrance hall, but no rooms were panelled and no furniture especially commissioned at the time of the opening.

Cedar's rarity has meant that it has been a wood seldom used by artist craftworkers in the craft revival of the late 20th century. Grant Vaughan's carved cedar bowl inspired by the sheath of the fruit of the *Monstera deliciosa* is a remarkable example of a contemporary Australian craftworker working in a timber that has fascinated Australian cabinetmakers, carpenters and craft workers since its discovery. It is also a rare example.

REFERENCE NOTES

INTRODUCTION

(pp 8–21)

1 In April 1802 Governor King issued an order forbidding the cutting of cedar without his permission, quoted in *Historical Records of Australia, Series 1*, vol 3, p619.

2 William Champ, 1808–92, arrived in Sydney as lieutenant with the 63rd regiment in 1828. After leaving the army he had a long, successful and varied career as a public servant and was the first premier of Tasmania. From 1857 to 1868 he was Inspector-General of penal establishments in Victoria and was largely responsible for the building of the Pentridge Penal Establishment.

3 Baron von Mueller, 1825–96, arrived in Australia from Germany in 1847. Appointed first Government Botanist in Melbourne in 1853 he was Director of the Botanic Gardens, Melbourne, from 1857 until 1873. A scholar and prodigious author, von Mueller was tireless in his promotion of the economic potential of Australian plants and timbers.

4 R T Baker, *Cabinet timbers of Australia*, New South Wales Government Printer, Sydney, 1913, p11.

5 Eugene von Guerard, 1811–1901, Austrian landscape painter, who worked in Australia from 1852 to 1882 and was appointed as the first Instructor of Painting in the School of Art and honorary curator at the Melbourne Gallery in 1870. Von Guerard painted numerous landscapes of eastern Australia and New Zealand which show him to be a keen observer of the subtle changes European occupation of the land had wrought. Von Guerard's *Australian landscapes*, a series of 24 tinted lithographs, is illustrative of the most striking and picturesque features of the landscape scenery of Victoria, NSW, South Australia, and Tasmania. Drawn from nature and lithographed by the artist with letter press description of each view, the series was published by Hamel and Ferguson, Melbourne, in 1866–67.

6 John Henderson, *Excursions and adventures in NSW with pictures of squatting and of life in the bush: an account of the climate, productions, and natural history of the colony, and of the manners and customs of the natives, with advice to emigrants etc* ... 2 vols. W Shoberl, London, 1851, pp 88–89.

7 ibid, pp122–130.

8 Henry Lawson, 1867–1922, author of poems and stories celebrating the characters of the Australian outback and the bush way of life. His work was first published in *The Bulletin* magazine which became a platform for the nationalism which led to Australian Federation in 1901.

9 Frederick McCubbin, 1855–1917, painter and teacher, a significant influence on the younger members of the Heidelberg School.

10 Tom Roberts, 1856–1931, painter and one of the most important members of the Heidelberg School.

11 James Broadbent, Suzanne Rickard & Margaret Steven, *India, China, Australia: trade and society 1788–1850*, Historic Houses Trust of NSW, Sydney, 2003.

12 The small size of cigar box panels meant that they were popular as a support for those artists wishing to paint in front of their subject. They did not require an easel and could be held in the hand or on the knees when painting and secured with pins and corks to transport works while still wet without paint being smudged or panels sticking together. For an account of outdoor painting in Australia, see Jane Clarke & Bridget *Whitelaw, Golden summers: Heidelberg and beyond*, National Gallery of Victoria, Melbourne, 1985.

13 Louis Abrahams, 1852–1903, painter associated with the Heidelberg School artists and businessman. He supplied his fellow artists with large numbers of cigar box lids obtained from his father's cigar factory in Melbourne.

14 Barnett Hyman Altson was a tobacconist in Elizabeth Street, Melbourne.

15 *9 x 5 impression* exhibition was held in August 1889 in Buxton's Galleries, Swanston Street, Melbourne, and introduced the new painterly landscapes of the young painters Charles Conder, Arthur Streeton, and Tom Roberts, accompanied by the older Frederick McCubbin, to an, at times, highly critical audience.

CHAPTER 1

(pp 22–41)

1 M L Brab. 7 A 78-6 f237; see *Historical Records of New South Wales* 5, p528.

2 In the United States, red cedar is the name for *Juniperus virginiana*, or *Thuja plicata*, both conifers.

D J Mabberley, *The plant-book*, (2nd edn, corr reprint), Cambridge University Press, 2002, p374.

3 The origin of the name mahogany is still debated, but it is likely that the word is a 17th-century European corruption of *oganow*, a name applied to the related African tree *Khaya ivorensis* by the Yoroba people of Nigeria and subsequently used by them for species of *Swietenia* in the Caribbean, where the Yoruba people were transported as slaves (Mabberley, *Plant-book*, p693). In Australia the name mahogany has in turn been transferred to *Eucalyptus robusta*, the swamp, or white, mahogany used at Sydney Cove by the First Fleet for furniture-making.

4 *Oxford English Dictionary*, 'cedar'.

5 D J Mabberley, 'Meliaceae', MS submitted to *Flora of Australia*, ABRS, Canberra.

6 D J Mabberley, 'Meliaceae', in M D Dassanayake (ed), *A revised handbook to the flora of Ceylon*, Amerind, New Delhi, 1995, pp 229–300

7 In Australia, the name rosewood is also used for the unrelated *Alectryon oleifolius* (boonaree, Sapindaceae) but elsewhere can refer to a wide range of timbers, especially those in the wattle family, Leguminosae, notably species of *Dalbergia* (palisander) and *Pterocarpus*, and so is also used in Australia for furniture made from the timber of these trees. See Mabberley, *Plant-book*, p623.

8 *Asiatick Researches* 4, 1795, p232.

9 R Desmond, *The European discovery of the Indian flora*, Oxford University Press, 1992, p41.

10 *Neue Schriften der Gesellschaft naturforschender Freunden zur Berlin* 4, 1803, pp180–224 esp p197; see K N Matthew, 'Notes on an important botanical trip (1799–1800) of J P Rottler on the Coromandel Coast (India) with a translation of his original text, explanatory notes and a map', *Botanical Journal of the Linnean Society*, 113, 1993, pp351–88, esp p358.

11 T Hardwicke, 'Enumeration of plants noticed in the preceding tour, between Hurdwar and Sirinagur, in the months of April and May 1796', *Asiatick Researches* 6, 1809, pp348–381 [p352].

12 See B T Styles, 'Sir William Jones' names of Indian Plants', *Taxon* 25, 1976, pp671–4; D J Mabberley, 'Francis Hamilton's commentaries with particular reference to Meliaceae', *Taxon* 26, 1977, pp523–40, esp p534. Jones used the Sanskrit name 'Cuveraca' in a system alternative to that of Linnaeus and that name is therefore unacceptable in modern nomenclature.

13 Desmond, p47.

14 B Henry, *British botanical and horticultural literature before 1800*, vol 2, Oxford University Press, 1975, pp189 et seq.

15 W Roxburgh †, *Flora Indica*, vol 3, Thacker & Co, Calcutta; Padbury, Allen & Co, London, 1832, p636.

16 J M Edmonds, '*Toona*', pp358–371, esp p370, in D J Mabberley et al, 'Meliaceae', *Flora Malesiana* 12, 1995, pp1–407.

17 Cf T G Vallance et al (compilers), *Nature's investigator. The diary of Robert Brown in Australia, 1801–1805*, ABRS, Canberra, 2001, p449; J Vader, *Red cedar: the tree of Australia's history*, Reed, Frenchs Forest, NSW, 1987, p18 has 'New Brush'. This was perhaps at North Brush Farm, now Eastwood, where up to at least 1986, there were still rainforest species typical of those associated with red cedar at nearby Brush Farm, though Toongabbie Creek (basalt) could have been the site (Doug Benson, Botanic Gardens Trust, Sydney, personal communication.)

18 *Historical Records of Australia, Series I*, vol 1, 1914, pp156–7.

19 Vader, p18.

20 *Historical Records of Australia, Series I*, vol 1, 1914, p308.

21 *Historical Records of Australia, Series I*, vol 2, 1914, p222; vol 3, 1915, p571; Vader, p21.

22 *Historical Records of Australia, Series I*, vol 3, 1915, p619.

23 ibid, p174.

24 *The narrative of a voyage of discovery, performed in His Majesty's Vessel the Lady Nelson ... to New South Wales*, Egerton, London, 1803.

25 *Historical Records of Australia, Series I*, vol 3, 1915, p257.

26 ibid, p258.

27 *Australian Dictionary of Biography*, Menzies.

28 ibid, Mary Reibey.

29 *Sydney Gazette*, 10 July 1803, 4c.

30 Vader, p35.

31 The surviving wings are now the NSW Parliament House and the Mint building in Macquarie Street.

32 Vader, p35.

33 British Library Add MSS 32 439 ff 41–2.

34 D J Mabberley, *Jupiter botanicus. Robert Brown of the British Museum*, Cramer, Braunschweig, 1985, pp40–43; J Webb, *George Caley Nineteenth[-]century naturalist*, Beatty & Sons, Chipping Norton, NSW, 1995; D J Mabberley, *Ferdinand Bauer: the nature of discovery*, Merrell Holberton and The Natural History Museum London, 1999, p84.

35 *Historical Records of Australia, Series I*, vol 5, 1915, pp619–20.

36 J E B Currey, *Reflections on the colony of New South Wales* [George Caley], Angus & Robertson, London, 1967, p106; *Australian Dictionary of Biography*, Mary Reibey.

37 *Australian Dictionary of Biography*, Menzies.

38 P Watts et al, *An exquisite eye: the Australian flora and fauna drawings 1801–1820 of Ferdinand Bauer*, Historic Houses Trust of NSW, Sydney, 1997, p152 + 160.

39 Webb, p49.

40 Vallance et al, *passim*.

41 Webb, p55. From as early as August 1801, Caley was writing to Banks, 'I mean to keep a bush native constant soon, as they can trace anything so well in the woods, and can climb trees with such ease, whereby they will be

very useful to me.'

42 Reibey's boat the *Raven* returned to Sydney from the Hawkesbury on 4 Nov 1803 (*Sydney Gazette*, 6 Nov 1803 1a), his boat the *James* returned from the Hunter on 11 Nov (*Sydney Gazette*, 13 Nov 1803 4b).

43 George Caley, Botanical Notebooks, B – Plants of Hunter River, f40 (The Natural History Museum London; microfilm FM4/2568, Mitchell Library, Sydney), transcribed from the microfilm by Caley's biographer, Joan Webb, and reproduced here with her kind permission.

44 Mabberley, *Jupiter botanicus*.

45 Webb, p47; Vallance et al, p529.

46 Vallance et al, p542.

47 ibid, p548.

48 ibid, p549.

49 ibid, pp550–1.

50 D J Mabberley & D T Moore, 'Catalogue of the holdings in the Natural History Museum (London) of the Australian botanical drawings of Ferdinand Bauer (1760–1826) and cognate materials relating to the *Investigator* voyage of 1801–1805', *Bulletin of the Natural History Botany Series* 29, 1999, p101, t22.

51 Brown correspondence [Brown MS 1, f277 v], Natural History Museum London, & British Library Add MSS 32439 f162 v; see *Historical Records of New South Wales*, 5, pp558–60 & Vallance et al, pp577–9.

52 *Familiarum naturalium regni vegetabilis. Synopses monographicae 1 (Hesperides)*. Landes-Industrie-Comptoirs, Weimar, pp131, 139. Note that 'autonyms' such as' '*Toona toona*' (which might have been expected in transferring *Cedrela toona* to the new genus), are not permitted in botanical nomenclature. Note also that the specific name has been applied to a completely unrelated tree from Queensland, the red siris or cedar acacia, a tree now called *Paraserianthes toona* (*Albizia toona*) in the wattle family (Leguminosae) because it has heartwood similar in appearance and uses to that of red cedar (Mabberley, *Plant-book*, p529).

53 E M Reid & M E J Chandler, *The London clay flora*, British Museum (Natural History), London, 1933, pp276–80.

54 *Records Geological Survey N.S. Wales*, 7, 1902, p4 + t15 fig 1.

55 E M Beekman, 'Introduction' in Georgius Everhardus Rumphius, *The Ambonese curiosity cabinet* (transl & ed by E M Beekman), Yale University Press, 1999, ppxxxv–cxii.

56 MS translation from G E Rumphius, *Herbarium amboinense* 3, 1743, and t39, in an appropriately dated style from the Dutch by Dr E M Beekman and reproduced here with his kind permission.

57 Edmonds, 'Toona', *Flora Malesiana* 12, 1995, pp358–371.

58 J M Edmonds & M Staniforth, 'Toona sinensis', *Curtis's Botanical Magazine*, ns 15, 1998, pp186–193 + t 348.

59 *Botanica, the illustrated A-Z of over 10,000 garden plants for Australian gardens and how to cultivate them*, Random House, Milsons Point, NSW, 1997, p892.

60 A G Floyd, *Rainforest trees of mainland south-eastern Australia*, Inkata Press, Melbourne & Sydney, 1989, pp207–8; D J Boland et al, *Forest trees of Australia* (4th edn), Nelson & CSIRO, Canberra, 1984, pp144–5.

61 Note that *Cedrela australis* was first used in 1829 as a replacement name for the valid name *C. toona* and, under the international rules of plant nomenclature, was a 'stillborn' name and therefore had to be discarded. *Toona australis*, dating from 1896, is a valid name but because of the rule of priority, would have to give way to one of a number of names such as *T. microcarpa* (C. DC.) Harms, a name based on a validly published earlier one, *Cedrela microcarpa* C. DC. (1878), if the Australian to Thailand plants were ever to be considered a species distinct from *T. ciliata*.

62 Edmonds, 'Toona', 1995; J M Edmonds, *The taxonomy and potential socio-economic importance of Toona species (Meliaceae)*, poster presented at Flora Malesiana Symposium, Royal Botanic Gardens, Kew, England, 1995. No acceptable name for such an entity as a subspecies has been published so far, though such plants have been treated by some authors as a variety, for which the correct name now appears to be *T. ciliata M.* Roem. var. *parviflora* (Benth.) Bahadur.

63 W G Keating & E Bolza, *Characteristics, properties and uses of timbers vol 1. South-east Asia, northern Australia and the Pacific*, Inkata Press, Melbourne & Sydney, 1982 no 343.

64 E Weber, *Invasive plant species of the world: a reference guide to environmental weeds*, CABI International, Wallingford, UK, 2003.

CHAPTER 2

(pp 42–59)

1 John Vader, *Red cedar: the tree of Australia's history*, Reed, Frenchs Forest, NSW, 1987, p21.

2 Kevin Frawley, 'The "inexhaustable scrub" and the closer settlement ideal: the northern rainforests in Queensland development', in *Queensland Geographical Journal*, 4th series, vol 3, 1988, p24.

3 ibid.

4 L J Webb, in A J Marshall (ed), *The great extermination*, Heineman, Melbourne,1966, p192.

5 A Harris, *Settlers and convicts, recollections of sixteen years' labour in the Australian backwoods*, Melbourne University Press, Melbourne, 1977, p31.

6 J Jervis, 'Cedar and the cedar-getters', *Journal and Proceedings of the Royal Australian Historical Society*, vol 25, no 2, Sydney 1940, p149.

7 Harris, p45.

8 C Hodgkinson, *Australia from Port Macquarie to Moreton Bay ...*, T & W Boone, London, 1845, p28.

9 ibid, p24.

10 Barron Field, *Geographical memoirs*

on New South Wales, John Murray, London, 1825, pp462–3.

11 Harris, p44.

12 Field, p462.

13 Hodgkinson, p32.

14 Eric Mjöberg, *Amongst stone age people in the Queensland wilderness*, Albert Bonniers, Sweden, 1918. Translated by Sivi M Fryer and edited by Rod Ritchie, p142

15 Mary Bundock, 'Notes on the Richmond River Blacks', 1898, Mitchell Library, State Library of New South Wales, Sydney, np.

16 Hodgkinson, p32.

17 W T Johnston & H A Tranter, *Christie Palmerston*, Eacham Historical Society, Herberton, 1982, p32.

18 A Meston, *Report on the Government Scientific Expedition to the Bellenden Ker Range*, James C Beal, Government Printer, Brisbane, 1889, p20.

19 Jervis, p149.

20 E E Morris (ed), *Cassell's Picturesque Australasia*, 4 vols, Melbourne, vol 4, p221.

CHAPTER 3

(pp 60–67)

1 Godfrey Mundy, *Our antipodes, or residence and rambles in the Australian colonies with a glimpse of the gold fields*, vol 3, Richard Bentley, London, 1852, pp10–11

2 Quoted from Francis Greenway's Specification Parramatta Parsonage, 15 April 1816.

3 Clement Hodgkinson, *Australia from Port Macquarie to Moreton Bay ...*, T & W Boone, London 1845, p12.

CHAPTER 4

(pp 68–81)

1 A G L Shaw (ed), *Gipps-La Trobe Correspondence*1839–1846. Letter 332, 21 June, 1845, Melbourne University Press, Melbourne, 1989, p337.

2 See B McGregor & Associates, *Government House Sydney Conservation and Management Plan*, Historic Houses Trust of NSW, Sydney, 1997.

3 F P Rollo Gillespie, *Viceregal Quarters: an account of the various residences of the Governors of New South Wales from 1788 until the present day*, Angus & Robertson, Sydney, 1975, p100.

4 *Specifications, New Government House, Sydney*, 20 March, 1834, Mitchell Library A1786. Cited in T Lane & J Serle, *Australians at Home: a documentary history of Australian domestic interiors from 1788 to 1914*, Oxford University Press, Melbourne, 1990, p75.

5 At the end of their terms of office various governors including Brisbane, Gipps, Denison and Robinson auctioned their surplus personal and household effects to recoup some of their establishment costs and to help defray the costs of re-locating back to Britain or to their next posting.

6 Sir George Gipps to Lord Russell, Mitchell Library A1267. Cited in B McGregor & Associates, p53.

7 K Fahy, C Simpson & A Simpson, *Nineteenth-century Australian furniture*, David Ell Press, Sydney, 1985, p538.

8 *Sydney Morning Herald*, 21 March, 1846, p2.

9 See illustration on p74.

10 J Kerr & J Broadbent, *Gothick taste in the colony of New South Wales*, David Ell Press in association with the Elizabeth Bay House Trust, Sydney, 1980, p118. First published in 1833, Loudon's book and other contemporary English pattern books and catalogues were the major sources of furniture designs for colonial cabinetmakers.

11 *Sydney Morning Herald* 19 April 1862, p9.

12 Fahy et al, *Nineteenth-century Australian furniture*, p41.

13 *Inventory of Furniture at Government House Sydney Enclosure with Lieutenant-General Darling's Despatch* dated 15th April 1831, 71 /3766 Public Record Office, pp 1/13. For inventory of imported mahogany furniture, see F P Rollo Gillespie, p xi.

14 To Colonial Secretary from Colonial Architect. Letter, 13 March 1847, not paginated. State Records Authority of NSW. 47/2208. Reference from Dr Rosemary Annable.

15 Government House owns the largest public collection of Andrew Lenehan provenanced furniture. One of his most significant commissions was a suite of cedar dining room furniture made for Sir William Denison in 1857. The history of this furniture is well documented in Fahy et al, *Nineteenth-century Australian furniture*, p53. Also *Votes and Proceedings of the NSW Legislative Assembly*, vol 5, 1862. Report from the Select Committee on the Petition of Mr Andrew Lenehan, Reference from K Fahy AM.

16 Government House Furniture Inventory (1846), State Records Authority of NSW. 21640, p5. The 1846 inventory lists the original ballroom furniture, as 'cedar seats or platforms, 15 carved cedar lustres for candles, 25 three-legged rough orchestra cedar stools and 9 rough cedar music stands', all essential for the hosting of formal entertainments such as balls and musical soirees. The word 'rough' is perhaps meant to convey that the quality or craftsmanship involved was coarser or less finished. Again, this furniture may have been made by the four artisans but can not be proven as none of the furniture has survived.

17 Government Architect's papers, *Estimate of Furniture and Stores Required for Government House as per Requisition 5. July 1866*, State Records Authority of NSW. 667 1450 (Archives Box 2/1450).

18 See R Conroy, 'James Cunningham woodcarver 1841–1903' in *Ashfield & District Historical Society Journal*, No 13, 2000 pp57–77. And J Wade, 'James Cunningham Sydney woodcarver' in *Australia Society Newsletter*, 1980/3 pp10–14.

19 Acquired at this time were a drawing room cabinet, a bonheur du jour and

a jardinere stand, all in rosewood and decorated with marquetry of diverse timbers. These are attributed to the Danish born cabinetmaker Christen Nielsen Lykke of Darlington.

20 *Australasian Builders' & Contractors' News*, 29 July, 1893, p54.

21 ibid, 9 December, 1893, p286.

22 Paul Greenhalgh, ed, *Art Nouveau 1890–1914*, Victoria & Albert Publications, London, 2000.

23 *Catalogue of New South Wales Exhibits. World's Columbian Exposition*, Chicago, 1893 (1894) pp417–419.

24 Report of the Executive Commissioner for New South Wales to the World's Columbian Exhibition, Chicago 1893. NSW Legislative Journals (1894/1895) vol 53, pp7–11.

25 Basil Burdett, 'Harold Parker' in *Art in Australia*, vol 1, No 2, May 1922, p32. Cited in J McKay, *Harold Parker sculptor*, Queensland Art Gallery, 1993, p44.

26 T Lane, *Robert Prenzel 1866–1941: his life and work*, National Gallery of Victoria, Melbourne, 1994.

27 *Building and Engineering Journal*, 11 June, 1892, p232.

28 See Fahy et al, *Nineteenth-century Australian furniture*, p59.

29 Fahy and Simpson have attributed the carved spiral decoration to the influence of Runic design, a link to Verdich's Scandinavian heritage. However the abovementioned article directly links the design of the Bartholomew suites to traditional New Zealand Maori culture. The idea of using indigenous cultures as a source for contemporary art and design was not novel. Ann Stephen in her study of the French-born artist Lucien Henry (1850–1896) points to the influence on Henry – a major advocate of the use of Australian flora and fauna in architectural design and decorative arts – of the South Pacific, in particular, Kanak designs.

30 In 1890–91 the Government Architect Walter Liberty Vernon designed the Arts and Crafts-inspired cottage The Chalet on the western terrace of Government House. I have now attributed the extant cedar furniture (hanging bookshelves and a pair of inglenook chairs) to the firm of George Bartholomew & Co on the basis of stylistic similarity to furniture supplied for other domestic commissions. See 'Art Furniture' in *Building and Engineering Journal*, 19 December 1891, p270. Cited in J McKay, p45.

31 The bookcase is illustrated in K Fahy, A Simpson, *Australian furniture: pictorial history and dictionary 1788–1938*, Casuarina Press, Sydney, 1998, p191.

32 D Ellesmore, *Government House Sydney, Survey of the Historic Photographs and References*. Notes for a Conservation Policy. Historic Buildings Group, Government Architects Branch, Public Works Department (1981).

CHAPTER 5

(pp 82–93)

1 NSW Parliamentary Debates, session 1883, First Series, vol 9, p1538.

2 City of Sydney Archives, CRS 65/1210/75.

3 'Bolection moulding, often ogee in section, covers, and projects beyond, the joint between two surfaces at different levels' in J A Gloag, *Short dictionary of furniture*, George Allen and Unwin, London, 1952.

4 City of Sydney CRS 65/1376/78.

5 City of Sydney CRS 65/1376/78.

6 City of Sydney CRS 179.

7 City of Sydney CRS 65/1210/75.

8 *Building and Engineering Journal*, 15 September 1888.

9 Particulars of Claim of William Coleman for 'Extras' in connection with his contract for completing staircase and room on first floor of Town Hall. CRS 21, 21/17 Mayors Reports 1885–8.

10 Centennial Hall Contract and Specifications CRS 65/1791.

11 Centennial Hall Contract and Specifications.

12 Works Committee, Orders of the Day, 22 December 1905.

13 Particulars of Claim of William Coleman for 'Extras' in connection with his contract for completing staircase and room on first floor of Town Hall. CRS 21, 21/17 Mayors Reports 1885–8.

14 In 1789, Captain Shea was buried on land he cultivated which officially became a burial ground in 1792. The Old Sydney Burial Ground closed in 1820 and the land was formally granted to the Municipal Council in 1869. The graves located in the area designated for the town hall foundations were relocated to Rookwood Necropolis in 1869.

CHAPTER 6

(pp 94–107)

1 G Butler Earp, *The gold colonies of Australia comprising their history, territorial divisions, produce and capabilities, how to get to the gold mines and every advice to emigrants*, Geo Routledge, London, 1852, p108.

2 Probably trees such as forest oak (*Allocasuarina torulosa*), swamp oak (*Casuarina glauca*) – both known as beefwood – red mahogany (*Eucalyptus resinifera*) and rosewood (*Dysoxylum fraserianum*).

3 William Bradley, A voyage to NSW. *The journal of William Bradley R N of HMS Sirius 1786–1792*, Public Library of NSW and Ure Smith, Sydney, 1969, p144.

4 Most likely red mahogany (*Eucalyptus resinifera*) which because of its hardness and deep red colour was sometimes compared with mahogany. Others likened *Toona ciliata* to mahogany because of its colouring. For an account of the earliest colonial cabinetmaking and the timbers used, see John Hawkins, 'The art of the cabinetmaker from the first settlement 1788 to 1820 (part 1) January–June 1983, (part 2) July–December 1983; and 'Art of the cabinet-maker 1788–1820: the known survivors', January–June 1984, in *The Australian Antique Collector*.

5 Thomas Sheraton, *The cabinet dictionary, containing an explanation of all the terms used in the cabinet, chair and upholstery branches, with directions for varnishing, polishing and gilding*, T Sheraton, London, 1803, pp88–90.

6 ibid.

7 See James Broadbent, Suzanne Rickard, & Margaret Steven, *India, China, Australia: trade and society 1788–1850*, Historic Houses Trust of NSW, Sydney, 2003.

8 See Hepplewhite & Co, *The cabinet-maker and upholsterer's guide ... from drawings*, London, 1788; Thomas Sheraton, *The cabinet-maker and upholsterer's drawing book*, I & J Taylor, London, 1791; and Thomas Shearer, *The cabinet-maker's book of prices and designs of cabinetwork*, London, 1793.

9 The King's Yard was situated on Sullivan's Cove, Hobart. The government lumber yard and furniture workshops was established in about 1819 and continued until 1835 when it came under the control of the Royal Engineers Department.

10 This secretaire desk is inscribed with the names of these two convicts who are thought to have been the cabinetmakers responsible for making it at the King's Yard, Hobart. See Kevin Fahy and Andrew Simpson, *Australian furniture: pictorial history and dictionary 1788–1938*, Casuarina Press, Sydney, 1998, pp134–135; 301.

11 Joseph Sly, c1802–1887, arrived in Sydney in 1834 and established a cabinetmaking workshop in Pitt Street. His firm was situated at various addresses in Pitt Street until it closed in 1861.

12 See John McPhee, *Two outstanding pieces in the Australiana collection*, Fine Arts Press, Sydney, 1990, pp49–51

13 J C Loudon, *An encyclopaedia of cottage, farm and villa architecture and furniture; containing numerous designs for dwelling, from the cottage to the villa, including farm houses, farmeries, and other agricultural buildings; several designs for country inns, public houses, and parochial schools; with the requisite firings-up, fixtures, and furniture; and appropriate offices, gardens, and garden scenery; each designed accompanied by analytical and critical remarks, illustrative of the principles of architectural science and taste on which it is composed*, Longman, Rees, Orme, Brown, Green & Longman, London, 1833, vol 2, p1090.

14 James Cunningham, 1841–1903, arrived from London in Sydney via New Zealand in 1863. He established a business specialising in architectural carving, as well as pattern-making, while he also carried out some cabinetmaking with heavily carved decoration. See John Wade & Robyn Conroy, 'James Cunningham, colonial woodcarver', *The Australian Antique Collector*, December 1997–June 1998, pp161–167.

15 For information about Australian Biedermeier, see Christopher Menz, *Colonial Biedermeier and German art in South Australia during the 19th century*, Art Gallery of South Australia, Adelaide, 1992.

16 See Graham Cornall, *Memories: a survey of early Australian furniture in the collection of the Lord McAlpine of West Green*, Australian City Properties Ltd, Perth, 1990.

17 Robert Prenzel, 1866–1941, arrived in Melbourne from Germany in 1888, the year of the Centennial International Exhibition. He established himself as a cabinetmaker and woodcarver developing a style that owed much to European art nouveau but was significant for its realistic use of Australian flora and fauna as carved decoration. See Terence Lane, *Robert Prenzel 1866–1941: his life and work*, National Gallery of Victoria, Melbourne, 1994.

18 Lucien Henry, 1850–1896, a French communard, was sentenced to transportation to New Caledonia in 1872. After arriving in Sydney in 1879, the year of the Sydney International Exhibition, he established himself as an artist and teacher. Promoting the idea of a national school of decorative arts and making many designs for objects and architectural projects he was disappointed in the lack of interest in his work and left for Europe to find a publisher for his designs. He died there and the work remained unpublished. The Powerhouse Museum, Sydney, mounted a major exhibition of his work. See *Visions of a Republic, the work of Lucien Henry: Paris, Noumea, Sydney*, Powerhouse Publishing, Sydney, 2001.

19 Richard Thomas Baker, 1852–1941, was an English teacher who arrived in Sydney in 1879 at the time of the Sydney International Exhibition. Baker worked as a teacher and was eventually appointed to the staff of the Technological, Industrial and Sanitary Museum, later the Museum of Applied Arts and Sciences, in Sydney. He was appointed director in 1898, a position he held until his retirement in 1921. His books celebrating Australian design, stone and timbers are typical expressions of the nationalism of the time.

20 Edward Wilhelm Verdich, c1855–1931, was born in Denmark and arrived in Queensland in about 1877. He established a workshop in Sydney in 1879, employing as many as 60 cabinetmakers in the 1890s. After his death his family continued the business as Celebrity Furniture.

21 See Glenn R Cooke, *Lady woodcarvers of Rockhampton*, Rockhampton Art Gallery, 2000, and Russell Atkinson, *Ellen Nora Payne: woodcarver of Tasmania*, R O Atkinson, Launceston, 1975.

22 John Vader, *Red cedar, the tree of Australia's history*, Reed, Frenchs Forest, NSW, 1987, pp137–138.

23 See Helaine Fendelman and Jonathan Taylor, *Tramp art: a folk art phenomenon*, Stewart, Tabori & Chang, New York, 1999.

24 R T Baker, *Cabinet timbers of*

REFERENCE NOTES CONTINUED

Australia, New South Wales Government Printer, Sydney, 1913, p11.

25 Schulim Krimper, 1893–1971, was born in Austria and arrived in Melbourne in 1939. He established himself as the pre-eminent Australian cabinetmaker in the modern style after the Second World War. A memorial exhibition of his work was held at the National Gallery of Victoria in 1975. See Mark Strizic and Terence Lane, *Schulim Krimper cabinetmaker*, Gryphon Books, Melbourne, 1987.

26 Fred Ward, 1900–1990, established himself as a furniture designer in the 1930s, and in the 1950s taught interior architecture at the University of Melbourne. He was responsible for the furnishing of several buildings of the Australian National University, in particular University House, in 1952. See *Fred Ward: a selection of furniture and drawings*, Australian National University, Canberra, 1996.

FURTHER READING

Baker, R T, *Cabinet timbers of Australia*, New South Wales Government Printer, Sydney, 1913.

Bootle, K R, *The commercial timbers of New South Wales*, Angus & Robertson, Sydney, 1971.

Broadbent, James, *The Australian colonial house. Architecture and society in New South Wales 1788–1842*, Hordern House in association with the Historic Houses Trust of New South Wales, Sydney, 1997.

Craig, Clifford, Fahy, Kevin & Robertson, E Graeme, *Early colonial furniture in New South Wales and Van Diemen's Land*, Georgian House, Melbourne, 1972.

Fahy, Kevin, Simpson, Christina & Simpson, Andrew, *Nineteenth-century Australian furniture*, David Ell Press, Sydney, 1985.

Fahy, Kevin & Simpson, Andrew, *Australian furniture: pictorial history and dictionary 1788–1938*, Casuarina Press, Sydney, 1998.

Floyd, A G, *Rainforest trees of mainland south-eastern Australia*, Inkata Press, Melbourne & Sydney, 1989.

Henderson, John, *Excursions and adventures in NSW with pictures of squatting and of life in the bush, an account of the climate, productions, and natural history of the colony, and of the manners and customs of the natives, with advice to emigrants etc*, 2 vols, W Shoberl, London, 1851.

Keating, W G & Bolza, E, *Characteristics, properties and uses of timber 1. South-east Asia, Northern Australia and the Pacific*, Inkata Press, Melbourne & Sydney, 1982.

Lane, Terence & Serle, Jessie, *Australians at home: a documentary history of Australian domestic interiors from 1788 to 1914*, Oxford University Press, Melbourne, 1990.

Mabberley, D J, *The plant-book*, 2nd edn, corr. reprint, Cambridge University Press, 2002.

Ritchie, Rod, *Seeing the rainforests in 19th-century Australia*, Rainforest Publishing, Paddington, NSW, 1989.

Vader, John, *Red cedar: the tree of Australia's history*, Reed Books, Frenchs Forest, 1987.

EXHIBITION CHECKLIST

ARCHITECTURAL FRAGMENTS AND JOINERY

1 *Twin spiral staircase*, scale 3 inches to 1 foot, 1890
J L Gardiner, Australia 1851–1929
red cedar, other Australian timbers, painted wood
188 x 90 cm diameter
Collection Powerhouse Museum, Sydney

2 *Column*, from the portico of Burdekin House, Macquarie Street, Sydney, 1841
attributed to James Hume
born Scotland c1797,
arrived Australia c1835, died 1868
painted red cedar and pine
290 x 55 cm diameter
National Trust of Australia (NSW)

3 *Fanlight*, from Subiaco, Rydalmere, NSW, 1836
John Verge, architect,
born England 1782,
arrived Australia 1828, died 1861
painted red cedar, glass
64 x 247 x 5 cm
Conservation Resource Centre, Historic Houses Trust

4 *Pilaster*, from Subiaco, Rydalmere, NSW, 1836
John Verge, architect
red cedar
291 x 32.5 x 7.5 cm
Conservation Resource Centre, Historic Houses Trust

5 *Gate*, from a balustrade, Tasmania, early 19th century
painted red cedar
91.5 x 95.5 x 6.5 cm
Private collection

6 *Chimneypiece surround*, Tasmania, early 19th century
maker unknown
red cedar
142 x 146 x 20 cm
Clive Lucas, Stapleton and Partners, Sydney

BOOKS

7 *Cabinet timbers of Australia*, R T Baker, Sydney, 1913
Private collection

8 *Cabinet timbers of Australia*, R T Baker, Sydney, 1913
Research Library, Historic Houses Trust

9 *Picturesque atlas of Australasia; illustrated under the supervision of Frederic B. Schell, assisted by leading colonial and American artists with over eight hundred engravings in wood.* Andrew Garran (ed), London, 1886
Research Library, Historic Houses Trust

10 *Excursions and adventures in NSW with pictures of squatting and of life in the bush, an acount of the climate, productions, and natural history of the colony, and of the manners and customs of natives, with advice to emigrants etc*, John Henderson, London, 1851
Private collection

11 *Australia from Port Macquarie to Moreton Bay: with description of the natives, their manners and customs; the geology, natural productions; fertility, and resources of that region; first explored and surveyed by order of the Colonial Government*, Clement Hodgkinson, London, 1845
Hordern House Rare Books, Sydney

12 *An encyclopaedia of cottage, farm and villa architecture and furniture; containing numerous designs for dwelling, from the cottage to the villa, including farm houses, farmeries, and other agricultural buildings; several designs for country inns, public houses, and parochial schools; with the requisite firings-up, fixtures, and furniture; and appropriate offices, gardens, and garden scenery; each design accompanied by analytical and critical remarks illustrative of the principles of architectural science and taste on which it is composed.* J C Loudon, vol 2, London, 183[illegible]
Private collection

13 *Amongst cannibals, an account of four years' travel in Australia and of camp life with the Aborigines of Queensland*, Carl Lumholtz, London, 1889
Private collection

14 *Fallander av ett storre urskogstrad, Malanda, N Queensland, in Bland stenaldersmanniskor I Queenslands vildmarker*, Eric Mjöberg, Stockholm, 1918
Private collection

15 *Our Antipodes: or residence and rambles in the Australasian colonies, with a glimpse of the gold fields*, Godfrey Charles Mundy, vol 3, London, 1852
Private collection

16 *Plants of the Coast of Coromandel*, William Roxburgh, vol 3, London, 1819 [1815–20]
Royal Botanic Garden, Edinburgh

17 *Herbarium Amboinense*, G E Rumphius, vol 3, Amsterdam, 1743
Botanic Gardens Trust Library, Sydney

18 *Agricultural Journal*, vol 1 part 2, Brisbane, August, 1897
John Oxley Library, State Library of Queensland

BOTANICAL SPECIMENS

19 *Toona ciliata*, collected on the Hunter River by Robert Brown, 22 October 1804
plant specimen mounted on paper, pen and ink, printed paper labels
44.5 x 26.2 cm
The Natural History Museum, London

20 *Toona ciliata*, collected in the Hunter Valley by George Caley, c1804
plant specimen mounted on paper, pen and ink, printed paper labels
44.5 x 26.2 cm
The Natural History Museum, London

FURNITURE AND DECORATIVE ARTS

21 *Salad box*, 1995
Fred Barrie, Australia 1903–1999
red cedar
8 x 25 cm diameter
inscribed on base: Fred S Barrie / 10-8-95
Private collection

22 *Royal arms of the United Kingdom*, c1865
James Cunningham, born England 1841, arrived Australia 1863, died 1903
red cedar, painted
86.5 x 123 x 15 cm
incised on reverse: J CUNNINGHAM
Government House, Historic Houses Trust

23 *Lectern*, c1873
attributed to James Cunningham
red cedar
144.5 x 54 x 67 cm
All Saints Anglican Cathedral, Bathurst

24 *Puzzle frame*, 1898
J A Dohnt, Shepparton, Victoria
photograph a replica of 1902 original
cedar, painted wood, glass
37.5 x 46 x 8.5 cm
Private collection

25 *Canopy and mayoral chair*, from the Council Chamber, Sydney Town Hall, 1883
attributed to John Hennessey, designer, born Ireland 1853, arrived Australia 1880, died 1924; William Coleman, Sydney, manufacturer
red cedar, leather upholstery
canopy: 325 x 201.5 x 69 cm, chair: 139 x 83 x 67 cm
Sydney Town Hall collection, City of Sydney

26 *Bench*, from the Council Chamber, Sydney Town Hall, 1883
attributed to John Hennessey, designer
William Coleman, manufacturer
red cedar, leather upholstery
129 x 439 x 59.2 cm
Sydney Town Hall collection, City of Sydney

27 *Tub chair*, from Council Chamber, Sydney Town Hall, 1883
attributed to John Hennessey, designer
William Coleman, manufacturer
red cedar, leather upholstery
83 x 74 x 62 cm
Sydney Town Hall collection, City of Sydney

28 *Picture frame*, c1865
attributed to Robin Lloyd Hood, Hobart, framemaker
born England 1828, arrived Australia 1833, died 1916
red cedar, blackwood, musk, glass, steel engraving
101.2 x 116 cm
Tasmanian Museum and Art Gallery, Hobart, Purchased with the assistance of The Art Foundation of Tasmania with funds raised from the Von Guerard 'Hobart Town' Appeal, 1996

29 *Tobacco jar*, c1948
Schulim Krimper, born Austria 1893, arrived Australia 1939, died 1971
red cedar, Queensland silver ash
24.8 x 8.9 cm diameter
National Gallery of Victoria, Melbourne
Gift of Schulim Krimper, 1949

30 *Chest of drawers on stand*, c1955
Schulim Krimper
red cedar, Queensland silver ash
119 x 104.5 x 46.2 cm
Art Gallery of South Australia, Adelaide, Ken Myer Fund 1992

31 *Pair of doors*, from an alderman's office, Sydney Town Hall, 1878
David McBeath, designer
Henry Smith and James Bennett, Woolloomooloo, builders
red cedar
both doors, 279.5 x 85.5 cm
Sydney Town Hall collection, City of Sydney

32 *Frame*, with chip-carved
decoration, c1900
T Porschoest, Australia
red cedar
53.4 x 42.8 cm
stamped on reverse:
T PORSCHOEST
Private collection

33 *Spores bowl*, 1982–83
Ben Trupperbaumer, born
Germany 1948, arrived Australia
1976
acacia cedar
10 x 75 cm diameter
incised on base: BEN T '82-83
National Gallery of Australia

34 *Bowl*, 1986
Grant Vaughan,
born Australia 1954
red cedar
14 x 25.5 x 22 cm
Collection Powerhouse Museum,
Sydney

35 *Secretaire cabinet*, c1892
E W Verdich,
born Denmark c1855,
arrived Australia 1855, died 1931
red cedar, beefwood, tulipwood,
brass hardware
266.5 x 248 x 68.5 cm
Government House,
Historic Houses Trust

36 *Chair*, from the Rare Books and
Manuscripts Reading Room,
National Library of Australia,
1969
Fred Ward, Australia 1900–1990
red cedar, replacement upholstery
84.5 x 56.5 x 53 cm
National Library of Australia,
Canberra

37 *D-end of dining table*,
Rouse Hill estate, NSW,
c1820
maker unknown
red cedar
72.5 x 130 x 65.5 cm
The Hamilton Rouse Hill Trust
Collection

38 *Side chair*, c1835
maker unknown, Sydney
red cedar,
cane (replacement of original)
89 x 58.5 x 53 cm
Private collection

39 *Hall chair*, 1846
attribted to John Arbuckle, Sydney
red cedar
129 x 56.5 x 50 cm
Government House,
Historic Houses Trust

40 *Miniature four-poster bed*, c1840
maker unknown, NSW
red cedar
46.5 x 33.3 x 50 cm
National Trust of Australia (NSW)

41 *Decorative roundel*, featuring a
kangaroo, late 19th century
maker unknown, Queensland
red cedar
30.5 cm diameter
Private collection

42 *Jewellery box*, inscribed with
The Lord's Prayer, 1912,
maker unknown, Barossa Valley,
South Australia
cedar, paint, cotton, mirror
18.5 x 27 x 27 cm
Private collection

43 *Pair of wine bottle holders*,
early 19th century
maker unknown, Tasmania
red cedar
10.5 x 10.5 x 18.5 cm
Private collection

44 *Side chair*, c1820
maker unknown, Tasmania
red cedar, replacement upholstery
81 x 45.5 x 38 cm
Private collection

45 *Workbox*, c1840
maker unknown, Tasmania
red cedar and various Tasmanian
woods on a pine base, glass,
mother-of-pearl, cotton
20.5 x 35.5 x 29.5 cm
Collection Powerhouse Museum,
Sydney

46 *Centre table*, c1840
maker unknown, Tasmania
red cedar, rosewood
75.5 x 125 cm diameter
Private collection

47 *Sofa*, c1840
maker unknown, Tasmania
red cedar, brass castors,
replacement horsehair upholstery
115 x 210 x 56 cm
National Gallery of Australia

48 *Miniature chest of drawers*, c1880
maker unknown, Tasmania
red cedar
34 x 33 x 16 cm
Private collection

49 *Miniature chest of drawers*, c1880
maker unknown, Tasmania
red cedar, porcelain
22.5 x 22 x 9.5 cm
Private collection

50 *Bookcase*, with 28 timber
specimens in the shape of books,
c1866
made by unidentified prisoner,
Pentridge Penal Establishment,
Melbourne
red cedar, and various other
Australian timbers
bookcase 57.5 x 51.2 x 16.7;
books each 17.5 x 11 x 2.5 cm
Tasmanian Museum and Art
Gallery, Hobart

51 *Puzzle frame*, c1900
maker unknown, Portland, Victoria
cedar, gelatin silver photograph,
glass, tin
31 x 23 x 5 cm
Private collection

52 *Photo frame*, c1900
maker unknown, Victoria
cedar, pine, painted wood
33.5 x 38 x 2 cm
Private collection

53 *Cigar box*, with chip-carved
decoration, early 20th century
maker unknown, Australia
unidentified timber
11 x 13.5 x 13.5 cm
Private collection

54 *Box*, with chip-carved decoration, early 20th century
maker unknown, Australia
red cedar
7.5 x 20 x 16 cm
Private collection

55 *Specimen cabinet*, made for Sir Joseph Banks's house at 14 New Burlington Street, London 1776
maker unknown, London
mahogany, ebony and unidentified timbers, brass, glass
56.5 x 56.5 x 39.5 cm
National Herbarium of New South Wales, Botanic Gardens Trust, Sydney

EPHEMERA

56 *Postcard*, Clearing scrub lands, from the *Pioneer Life* series, c1906
Charles Kerry, photographer, Australia 1858–1928
photo lithograph, with printed colour
8.8 x 14 cm
Private collection

57 *Postcard*, A settler's hut, Lismore, c1910
McKenzie, photographer; Syer and Co, publishers
photo lithograph
8.8 x 13.9 cm
Private collection

58 *Postcard*, Mount Tambourine, Queensland, from the *Queensland Views* series, c1910
printed in Germany
photo lithograph, with printed colour
13.9 x 8.9 cm
Private collection

59 *Postcard*, Road scene, Camberwarra Mountain, Shoalhaven district, from the *Nature* series, c1910
printed in Germany
photo lithograph, with printed colour
8.9 x 13.9 cm
Private collection

60 *Postcard*, A peep on the Bulli Pass, Illawarra, from the *Nature* series, c1910
printed in Germany
photo lithograph, with printed colour
8.8 x 13.9 cm
Private collection

61 *Postcard*, Camberwarra Pass, from the *Nature* series, c1910
printed in Germany
photo lithograph, with printed colour
8.8 x 14 cm
Private collection

62 *Postcard*, The Big Scrub, Richmond River, c1910
printed in Germany
photo lithograph, with printed colour
9 x 13.9 cm
Private collection

63 *Postcard*, Cabbage trees, Illawarra, from the *Nature* series, c1910
printed in Germany
photo lithograph, with printed colour
9 x 14 cm
Private collection

HISTORICAL ARTEFACTS

64 *Single-scull rowing shell, and blades*, c1910
James Edwards, boat builder, Melbourne, active c1870–1925
red cedar, unidentified hardwoods, aluminium alloy, metals, parachute material, leather shoes, rubber
scull 730 x 150 x 20 cm; blades 284 x 16.5 x 10 cm
stamped in ink on seat: J EDWARDS MAKER MELBOURNE
Newcastle Regional Museum

65 *Theodolite transit box*, 1855–1865
T F Wiesener, Astronomical, Mathematical, Optical, Surveying and Philosophical Instrument Maker, Steam Factory and Shop, George Street, Sydney, 1871–1918
red cedar
25 x 39 x 27 cm
Printed paper label for T Wiesener
Collection Powerhouse Museum, Sydney

66 *Grave board*, 1866
maker unknown, Armidale, NSW
red cedar
130 x 61 x 9 cm
inscription: SACRED / TO THE MEMORY OF / Obed Roan / who departed this life 30th April 1866 / in the 15th year of his life / Our life is ever on the wing / And death is ever nigh / The moment that our lives begin / We all begin to die.
Armidale Folk Museum, Armidale and Dumaresq Council

67 *Coffin*, pre-1840
maker unknown, Sydney
red cedar
45 x 190 x 55 cm approx
St Andrew's Cathedral, Sydney

68 *Box for troy weight set*, c1883
maker unknown, Sydney
red cedar, glass, cotton
11 x 25.5 x 19 cm
Collection Powerhouse Museum, Sydney, Gift of J L McAlister, 1917

69 *Photographer's gyroscope*, c1880
maker unknown, NSW
red cedar
123 x 42 x 44 cm
National Trust of Australia (NSW)

70 *Electra*, 30" tonner, pond yacht on stand, 1880–1900
maker unknown, NSW
red cedar, kauri, Huon pine, western red cedar, cotton, brass
142 x 123 x 19.5 cm
Collection Jack Woodward, Model Shipyard, Balmain

71 *Whet-stone box*, c1900,
maker unknown, NSW
red cedar, whet-stone
6 x 23.7 x 6.2 cm
Private collection

72 *Evans Patent Combination Van*,
model railway carriage, c1900
maker unknown, NSW
red cedar, pine, nickel plated metal
and cotton
50.5 x 85 x 38 cm
Private collection

MANUSCRIPTS

73 *Diary*, entry noting the collection
of red cedar specimen,
22 October 1804
Robert Brown, born Scotland
1773, died England 1858,
worked in Australia 1801–05
pen and ink on paper
20 x 23.5 cm
The Natural History Museum,
London

74 *Letter*, Robert Brown to Sir Joseph
Banks discussing red cedar,
21 February 1805
pen and ink on paper
24.4 x 19.4 cm and 20.2 x 15.4 cm
The Natural History Museum,
London

75 *Botanical notebook*,
written at Parramatta, March 1804
George Caley,
England, 1770–1829,
worked in Australia 1800–09
pen and ink on paper
16.5 x 10.5 cm
The Natural History Museum,
London

76 *A timber getter's contract*, 1835
Michael Chey and William
Holland
pen and ink on paper
31.3 x 20 irreg cm
Private collection

PHOTOGRAPHS

77 *A cedar tree*, Gloucester, NSW,
1944
Fred Barrie, Australia 1903–1999
photographic dye, gelatin silver
photograph in red cedar frame
frame 120.5 x 75 cm
photograph 107.5 x 62 cm (sight)
Newcastle Regional Museum

78 *Felling a cedar tree, Gloucester,*
NSW, 1944
Fred Barrie
photographic dye, gelatin silver
photograph in red cedar frame
frame 120.5 x 77.5 cm
photograph 107.5 x 64.5 cm
(sight)
Newcastle Regional Museum

79 *Hutten der cederholzfaller*
(Blackall Range), in an album
presented to the German Consul J
C Heussler, 1897
J J Hogg, photographer, Brisbane,
Queensland, active 1886–c1900
albumen silver photograph in
album with watercolour
decorations and silver mounts
image 15.2 x 22 cm
album page 28 x 39 cm
John Oxley Library, State Library
of Queensland

80 *Cedar log*, 1880
Charles Kerry & Co, Sydney,
Australia 1858–1928
albumen paper photograph
15.3 x 20.3 cm
Josef Lebovic Gallery

81 *Felling a forest veteran*, 1880s
Charles Kerry & Co
albumen paper photograph
20.1 x 15.3 cm
inscribed l.r.: KERRY PHOTO
Josef Lebovic Gallery

82 *Man felling tree using a spring-*
board, 1880s
attributed to Charles Kerry & Co
albumen paper photograph
15.1 x 20.1cm
Josef Lebovic Gallery

83 *Paddington/Woollahra steam tram*
no 80 passing Metropole Hotel,
Sydney, c1900
photographer unknown, Sydney
albumen silver photograph
20.2 x 15.4 cm
Josef Lebovic Gallery

84 *Cedar (Cedrela toona), portion of*
a giant tree – age 350 years 1924
photographer unknown, Sydney
gelatin silver photograph
37.3 x 46.7 cm image
Collection Powerhouse Museum,
Sydney

85 *Team on Bulli Road*,
late 19th century
photographer unknown, NSW
albumen silver photograph
15.3 x 20.2 cm
Mitchell Library,
State Library of NSW

86 *Bowman township, near*
Gloucester, NSW, c1900
photographer unknown, NSW
print-out paper photograph
15.2 x 21cm
Josef Lebovic Gallery

87 *Bulli, timber cutters*, c1900
photographer unknown, NSW
albumen silver photograph
image 13.9 x 10 cm
mount 16.3 x 10.8 cm
Mitchell Library,
State Library of NSW

TIMBER SPECIMENS

88 *Board from a cedar log*, exhibited
at Melbourne Centennial
Exhibition, 1888, NSW
red cedar
105 x 315 x 3 cm
painted: Cedar / Cedrela Toona
Roxb. / (Meliacene) / North Coast
Districts NSW
Collection Powerhouse Museum
Sydney

89 *Cross section of a cedar log*, late 19th century, NSW
red cedar
97.5 x 110 x 4.5 cm
painted: Red cedar / *Cedrela australis* TUM / Meliaceae
Collection Powerhouse Museum, Sydney. Purchased 1891

90 *Timber specimen*, turned column, 1900–1930, Sydney
red cedar, paint
104 x 42 cm diameter
painted: Red cedar / *Cedrela toona*, Roxb / Meliaceae
Collection Powerhouse Museum, Sydney

TOOLS

91 *Springboard*, 20th century
maker unknown, Port Macquarie, NSW
coachwood, iron
181 x 20 cm
Timbertown Collection, Hastings Council

92 *Crosscut saw*, c1825
William Sykes and Co, Sheffield, England
steel, wood (unidentified)
23 x 132 cm
stamped: William Sykes & Co / Sheffield
Private collection

93 *Pit saw*, mid 19th-century
maker unknown, England
steel
199.5 x 28.5 cm
Private collection

94 *Maul*, 19th-century
maker unknown, Australia
eucalypt (unidentified), steel
67.5 x 14.5 cm diameter
Private collection

95 *Log chain*, c1900
maker unknown, Australia
steel
148 cm length
Private collection

VIDEO

96 *Red cedar: felling and milling by the Carson Northern Timber Co Ltd, Gloucester*, 1944,
filmed by Fred Barrie
colour 16 mm film
Lent by the Barrie family

97 *Red gold*, 1963,
directed by Robert McAuley
colour 16 mm film
Courtesy Haydon family
Collection ScreenSound Australia, The National Screen and Sound Archive

WORKS OF ART

98 *A room in Woroongundi; a squatter taking it easy*, c1847
attributed to Charles Archer
born Scotland 1813,
died England 1862,
worked Australia 1841–1857
watercolour and gouache
11 x 16 cm (sight)
John Oxley Library, State Library of Queensland

99 *Toona ciliata*, c1805–10
Ferdinand Bauer
born Valtice (Czech Republic) 1760, died Vienna 1826,
worked Australia 1801–05
pencil and watercolour
52.7 x 35.9 cm
The Natural History Museum, London

100 *Hunting the kangaroo, from Field Sports ... of the native inhabitants of New South Wales*, London, 1813
Mathew Dubourg, engraver, England, active 1808–38,
after John Heaviside Clark, England 1770–1863
hand coloured etching and aquatint
14.4 x 19.4 cm printed image
Gregg collection

101 *Smoking out the opossum, from Field Sports ... of the native inhabitants of New South Wales*, London, 1813
Mathew Dubourg,
after John Heaviside Clark
hand coloured etching and aquatint on paper
19.5 x 14.5 cm printed image
Gregg collection

102 *Robert Sheppeard*, c1853
Edmund Edgar,
born England c1800,
arrived Australia 1826, died?
watercolour
37.5 x 27.5 cm oval (sight)
Private collection

103 *Sarah and Hepsabah Sheppeard*, c1853
Edmund Edgar
watercolour
37.5 x 27 cm oval (sight)
Private collection

104 *Still life*, 1840
W B Gould, born England 1803,
arrived Australia 1827, died 1853
oil on panel (red cedar),
pine (unidentified) frame
panel 22.7 x 19.8 cm,
frame 33.7 x 31 cm
painted inscription l.l.: Painted by WB Gould / 1840 inscribed on paper label on reverse: Painted by / WB Gould / 1840 / Framed by / Robin Hood / Hobart Town / Van Diemen's Land
Private collection

105 *Kiama, Illawarra*, NSW, 1869
after a sketch made in 1830
Henry Gritten,
born England 1818,
arrived Australia 1853, died 1873
after Robert Hoddle,
born England 1794,
arrived Australia 1823, died 1881
oil on canvas
58.5 x 45.5
National Library of Australia, Canberra

106 *Cabbage tree forest, American Creek* (N.S.W.), plate no. 15, *Australian landscapes*, Melbourne, 1866–68
Eugene von Guerard, born Austria 1811, died England 1901
worked Australia 1852–1881
colour lithograph
33.4 x 53 cm image
printed within image l.l.:
Eug. V. Guerard
Wollongong City Gallery
The George and Nerissa Johnson Memorial Bequest, purchased 1992

107 *Cabbage Tree Forest, American Creek, New South Wales*, 1860
Eugene von Guerard
oil on canvas
51 x 85.5 cm
inscribed l.l.: Eug: v : Guerard 1860
Wollongong City Gallery
Acquired with assistance from the Wollongong Gallery Society, NSW Office of the Minister for the Arts, and Public Subscription, 1984

108 *View of Sydney Cove, NSW*, 1804
Francis Jukes,
England 1747–1812, engraver
after Edward Dayes,
England 1763–1804
after Thomas Watling,
born Scotland 1762,
arrived Australia 1792, died 1801
hand coloured aquatint
39.5 x 60.5 cm image
The Beat Knoblauch Collection

109 *Sketchbook*, with a drawing of *Drue Walla*, Illawarra, 1840–44
Abraham Lincoln
born England 1815
arrived Australia 1838, died 1884
pencil on paper
page size 38 x 26 cm,
album 39 x 28 cm
inscribed in pencil: Drue Walla / Illawarra / The residence of my friend R.B. Fry Esq.re an Irish Gentleman. / This little Farmhouse is built almost entirely of cedar, and / stands most pleasantly on a little mountain stream, a tri / butary of the Minamurra River
Mitchell Library,
State Library of NSW

110 *Brush scene, Brisbane Water*, 1848
Conrad Martens,
born England 1801,
arrived Australia 1835, died 1878
oil on canvas
75 x 63 cm
inscribed l.l.: C Martens 1848
Dixson Galleries,
State Library of NSW

111 *Josias Hancock's saw-mills, Ipswich*, c1888
Phillip-Stephan Photo-Litho and Typographic Process Co Ltd
Woolloomooloo, NSW, 1884–1910
coloured photo-lithograph
image 14.2 x 21.5 cm
paper 21.7 x 27.5 cm
inscribed on margin l.l.: Phillip-Stephan Photo-Litho Process Co. Sydney; l.c.: JOSIAS HANCOCK'S SAW-MILLS / Ipswich
Josef Lebovic Gallery

112 *Silent Bob Bates*, 1895
Tom Roberts, born England 1856
arrived Australia 1869, died 1931
oil on panel (red cedar)
76.4 x 39.5 cm
inscribed l.c.: T ROBERTS / AP 95, l.c.: SILENT / BOB / BATES
Art Gallery of New South Wales, Purchased 1947

113 *Cedrela toona*,
drawn for William Roxburgh,
artist unknown, India
pencil, pen and ink and watercolour
33.5 x 46 cm
Royal Botanic Gardens, Kew

INDEX

Page numbers in bold print refer to main entries. Page numbers in italics refer to illustrations, photographs and paintings

ACKNOWLEDGMENTS

An exhibition and publication dedicated to a tree and its timber is unusual but not unique. In 1977 the Tasmanian Museum and Art Gallery mounted a small exhibition about the Huon pine, Tasmania's endemic, almost mythological, tree. Through a combination of botanical, historical and fine art scholarship, the exhibition traced the tree's discovery and illustrated its uses in such diverse areas as boat building, furniture and sculpture. It is my memory of this fascinating exhibition that planted the seed for an exhibition and publication devoted to Australian red cedar.

Two existing publications have provided invaluable references and inspiration in the preparation of this exhibition and catalogue: John Vader's *Red cedar, the tree of Australia's history* (1987) which focuses on the important role of cedar in the history and economic development of NSW and Queensland, and Rod Ritchie's, *Seeing the rainforests in 19th-century Australia*, (1989). These books continue to be important sources of information on the history of the timber industry and our use and perception of Australia's natural resources and landscape.

Of course, the exhibition and publication would not have occurred without the enthusiasm and support of Peter Watts, the Director of the Historic Houses Trust of NSW and its dedicated staff who in many and various ways have ensured that everything happened – on time!

Numerous curatorial colleagues and friends have put up with strange requests for information, and in many instances been able to supply it. In particular I wish to thank the following institutions and their staff: All Saints Anglican Cathedral, Bathurst: Richard Hurford, Bishop of Bathurst, Andrew Sempell, Dean of Bathurst; Armidale Folk Museum: Peter Chambers, Curator; Art Gallery of New South Wales: Edmund Capon, Director; Art Gallery of South Australia: Ron Radford, Director, Robert Reason, Associate Curator of Decorative Arts; Centre for Cultural Materials Conservation, University of Melbourne: Robyn Sloggett, Director, Lucy Harper, APA(I) Researcher; City of Sydney: Margaret Betteridge, Curator, Sydney Town Hall Collection; The Hamilton Rouse Hill Trust Collection; Macleay Museum, University of Sydney: Geoff Barker, Curator of Historic Photographs; National Gallery of Australia: Brian Kennedy, Director, Robert Bell, Curator of Decorative Arts, Gael Newton, Curator of Photography, Roger Butler, Curator of Australian Prints and Drawings, Margaret Shaw, Chief Librarian, Gillian Currie, Deputy Chief Librarian; National Gallery of Victoria: Gerard Vaughan, Director, Isobel Crombie, Curator of Photography, Terence Lane, Senior Curator, Australian Art to 1900, Jennifer Phipps, Curator of Australian Art, Late Modernism; National Library of Australia: Jan Fullarton, Director General, Greta North, Registrar of Exhibitions, Sylvia Carr, Pictures Reference Librarian; National Trust of Australia (NSW): Elsa Atkin, Executive Director, Patricia McDonald, Curator; Natural History Museum, London: Christopher Mills, Head of Library Services, R Bateman, Keeper of Botany; Newcastle Region Art Gallery: Nick Mitzevich, Director, Katrina Rumley, Curator; Newcastle Region Museum: Gavin Fry, Director, Julie Baird, Curator; Powerhouse Museum, Sydney: Kevin Fewster, Director, Anne Watson, Curator of Decorative Arts and Design, Michael Lea, Curator, Music and Musical Instruments; Parliament House, Canberra: Louise Dauth, Director, Parliament House Art Collection; Queen Victoria Museum and Art Gallery, Launceston: Chris Tassell, Director, Bridget Arkless, Curator of Fine Art, Glenda King, Curator of Craft, Rhonda Hamilton, Curator of Community History; Royal Botanic Gardens, Edinburgh: Jane Hutcheon, Head of Library; Royal Botanic Gardens, Kew, England: Peter Crane, Director; Botanic Gardens Trust, Sydney: Frank Howarth, Director, David Mabberley, Honorary Research Associate; St Andrews Cathedral: Phillip Jensen, Dean of Sydney; State Library of New South Wales: Dagmar Schmidmaier, State Librarian, Elizabeth Ellis, Assistant State Librarian, Collection Management Services and Mitchell Librarian, Richard Neville, Manager Original Materials Branch, Alan Davies,

Curator of Photographs; State Library of Queensland: Brian Randall, Manager, Diane Byrne, Reference Librarian, Angelo Comino, Administrative Officer, Pictorial Collection; Tasmanian Museum and Art Gallery: Bill Bleathman, Director, David Hansen, Curator of Fine Art, Peter Hughes, Curator of Decorative Art, Erica Burgess, Conservator; Victoria University of Wellington, New Zealand: Roger Blackley, Senior Lecturer; Wollongong City Gallery: Peter O'Neill, Director, Louise Brand, Curator.

Individuals who have offered assistance and advice include: Kim Akerman, Julian Bickersteth, International Conservation Services, John Blaine, Michael Bogle, Tim Bonyhady, Australian National University, James Broadbent, Caroline Butler-Bowden, John Chapman, Lauraine Diggins, Lauraine Diggins Fine Art, Tim McCormick, Australian Arts, Penny Coleing, Steve Crewes, John Denton, Pat Flinn, Mary McKillop Penola Centre, Liz Gillroy, Irene and Robert Goard, Robert Holden, Janie and Theo Hooy, Jo Kamminga, Khai Liew, Augusta Antiques, Jac MacNaughtan, Bebe and Bob MacPherson, Judith McKay, Michael McWilliam, Longford Antiques, Megan Martin, Edgar Penzig, Rod Ritchie, Andrew Simpson, Ben Trupperbaumer, John Wade and Pat Wallace.

Most importantly I wish to thank those individuals who have lent works to this exhibition, including: Alicia Barrie, Toni and Peter Bessant, Margaret and Jim Bleasel, Trish and Rod Crockford, Kevin Fahy, Ted Gregg, Viki and Roger Harrison, Ralph Hawkins, Beat Knoblauch, Keith Okey, Josef Lebovic, Josef Lebovic Gallery, Clive Lucas, Lucas Stapleton and Partners, Megan Martin, Anne McCormick, Hordern House Rare Books, Dick Phillips, Rod Ritchie, Pamela Rowell, Jim Sait, Urs Schwarzenbach, Jennifer I Stuerzl and David Bedford, and Jack Woodward, Model Shipyard.

Both private and institutional lenders have been generous, often in the extreme – and I wonder what some private collectors are sitting on or eating off! Treasured objects have been lent so that they might be shared with a wider audience. No greater testimony to their generosity could be imagined.

JOHN MCPHEE
VISITING CURATOR